Essential Elements for a GMP Analytical Chemistry Department

Thomas Catalano

Essential Elements for a GMP Analytical Chemistry Department

Springer

Thomas Catalano Ph.D.
PharmChem Analytical Consultants LLC
Buffalo Grove, IL
USA

ISBN 978-1-4614-7641-2 ISBN 978-1-4614-7642-9 (eBook)
DOI 10.1007/978-1-4614-7642-9
Springer New York Heidelberg Dordrecht London

Library of Congress Control Number: 2013938176

Printed on acid-free paper

Springer is part of Springer Science+Business Media (www.springer.com)

This Book is dedicated to my Wife Jeanie
Who helped with the Manuscript
And is A Lifelong Friend

Preface

This volume presents a systematic approach to understanding the essential elements required for a successful GMP analytical chemistry department to function as an efficient and effective organization. It describes in detail a department structure which allows for the necessary processes to become available to all its personnel. This department structure facilitates the free flow of information and interactions within the department. The environment and culture created by this approach encourages and rewards the sharing of ideas, skills, and abilities among department personnel.

The essential elements such as SOPs, regulatory guidances/guidelines, project teams, technical and department processes, statistical concepts, outsourcing, mentoring and hiring the best are among the many topics that are discussed in detail and how they can be implemented to build an efficient and effective analytical department.

Essential Eelements for a GMP Analytical Chemistry Department can be an asset to all companies that perform GMP analytical method development, validation, analyses, etc. This book is an essential addition to the library for all start-ups, generic pharmaceutical companies, and contract research organizations. In addition, it can be an important subject as part of a university Pharmaceutical Technology program.

Thomas Catalano, Ph.D.
Buffalo Grove, IL
USA

Contents

Abbreviations

$\%R_c$	Percent risk change between plans
$\%R_t$	Percent total risk of current plan
$\%R_{XL}$	Percent risk evaluation of determined low risk plan
ACN	Acetonitrile
AD	Analytical Department
AL	Analytical Lead
API	Active Pharmaceutical Ingredient
BA	Bioavailability
BCS	Biological classification system
BE	Bioequivalence
CDA	Confidentiality agreement
CFR	Code of federal regulations
CI	Confidence interval
CMC	Chemistry manufacturing and controls
CoA	Certificate of analysis
CTD	Common technical document
CTM	Clinical trial material
DAD	Diode array detector
DC	Development compound
Deg. Prds	Degradation products
Disc	Discovery
DMPK	Drug metabolism pharmacokinetics
DOE	Design of experiments
DSC	Differential scanning calorimetry
EC	Electron capture detector
FID	Flame ionization detector
GC	Gas chromatography
GLP	Good laboratory practices
GMP	Good manufacturing practices
HLA	Highest level of authority
HPLC	High performance liquid chromatography
ICH	International conference on harmonization
ID	Identity, identification

IMP	Impurity
IND	Investigative new drug
IP	Intellectual property
IQ	Installation qualification
IR-ID	Infra-Red spectroscopy identification
IVIVC	Invitro/Invivo correlations
k'	Capacity factor
LC/MS	Liquid chromatography/Mass spectrometry
LIMS	Laboratory information management system
LOD	Limit of detection
LOQ	Limit of quantitation
MeOH	Methyl alcohol
MSDS	Material safety data sheet
NDA	New drug application
NMR	Nuclear magnetic resonance
NPD	Nitrogen phosphorous detector
OQ	Operational qualification
PH	Phase
PK	Pharmacokinetic
PQ	Performance qualification
QA	Quality assurance
Ref. Std	Reference standard
RFI	Request for information
ROA	Report of analysis
RSD	Relative standard deviations
RSM	Reference standard manager
SD	Standard deviations
SOP	Standard operating procedure
THF	Tetrahydrofuran
Tox.	Toxicology
USP	United States pharmacopeia
x_c	Total risk value for current plan
x_L	Total risk value for determined low risk plan
x_{max}	Maximum total risk value
x_{min}	Minimum total risk value
x_n	Total risk value for new plan
XRD	X-Ray diffraction

Chapter 1
Introduction

After working for many years in various analytical departments it has become apparent what essential elements are required for a successful GMP Analytical Chemistry Department to function as an efficient and effective organization. The structure of the department needs have the necessary processes available to all its personnel and the department must be structured in a way where there is a free flow of information and interaction. The isolation and hording of ideas, skills and abilities should be discouraged and the sharing and helping each other should be rewarded.

Among the most basic essentials, and a requirement, in any successful GMP Analytical Chemistry Department is the establishment of Standard Operation Procedures (SOP's) and working Guidelines. This is the most efficient way to communicate best practices within the group and establish minimum standard requirements which must be met. It is a requirement and a good practice that personnel are trained on these SOP's and guidance's annually. In addition it is a good practice that analytical personnel are made familiar with the contents of regulatory documents, such as the CMC, so that it allows the staff to understand the need for the attention to detail and the stringent documentation requirements placed on them.

Processes should be established over a range of disciplines needed for the department to effectively apply its technology to the challenges which will be forthcoming. In order for processes to be adhered to, a process owner or, in larger organizations, expert groups should be created for the project teams to utilize when appropriate. Project teams are the backbone of the organization. The department should have a specified process for the formation of project teams, so that the teams have proper representation. Proper functioning of project teams is essential for the success of the project to be accomplished. The team goes through several stages of development; Forming, Norming, Storming, and Preforming before it will become a full functioning team. In order to manage the team effectively a process should be followed which will describe all of the team dynamics and individual responsibilities that are necessary for success. Knowledge of statistics is very important for analytical chemist to interpret the analytical data. The chapter

T. Catalano, *Essential Elements for a GMP Analytical Chemistry Department*,
DOI: 10.1007/978-1-4614-7642-9_1,

on statistical concepts was included to identify and briefly review those concepts such as, significance, confidence intervals, uncertainty in measurements, design of experiments, etc. that are critical for the analytical chemist to understand so that they can discuss the data with company and regulatory agency statisticians and ensure that the data interpretation is acceptable for its intended use.

Another important element for an effective Analytical Department is the motivation of personnel. It may be hard to believe, but compensation is *Not* the most motivating factor; however it is among the top three. Job satisfaction and professional position in the organization appear to be among the highest motivators. A defined Career path is an excellent way to motivate personnel, so that each individual knows what they must accomplish to climb the career ladder. It also allows for lateral movement within the department so that the department does not become top heavy by utilizing promotions as the only avenue for growth. Another element for motivation is to consider a mentoring program. This is very well received, especially by younger staff members, because it allows the opportunity for professional growth without having to wait for the opportunity to come across their path during the regular course of business.

In most departments there is always a need to outsource project activities for various reasons. However, these reasons should be thoroughly rationalized as to the future impact on the project. One aspect of outsourcing is to be certain of the overall quality of the organization being utilized, thus having a comprehensive outsourcing process will ensure that all aspects are being addressed and nothing is overlooked. Generally speaking most departments will try to outsource more of the routine work such as, well understood methods and activities, while keeping the core competencies in house.

Hiring is an essential element within the Analytical Department to keep the department growing with new talent and replacing talent that has left the organization. There is a fierce competition for good talent among the various companies. A hiring process should be in place which is ongoing and is constantly connected with colleges, universities and recruiters. Implementing searches for key talent only when positions are available usually results in settling rather than selecting.

The following chapters in this book will describe in detail the essential elements introduced above and how they can be implemented to build an efficient and effective Analytical Department.

Chapter 2
Organization

Abstract In this chapter, an organizational structure is proposed which will allow for a more efficient and effective Analytical Department. The organization is based on a project team model. Discussions on how the team is formed, managed, and the utilization of team dynamics are presented in detail. A process for project risk evaluation based on the project plan is proposed. Among the most basic essentials, and a requirement, in any successful GMP Analytical Chemistry Department is the establishment of Standard Operation Procedures (SOP's) and working Guidelines. This is the most efficient way to communicate best practices within the group. The essential SOP's that should be included within a GMP Analytical Chemistry Department is listed within the chapter, along with a template for writing SOP's.

2.1 Department Structure

The Organization of an analytical department has a large impact on its efficiency. Having a good exchange of information and good interaction between staff personnel is extremely important. The organization structure is the starting point for creating an effective and efficient analytical department. The structure should contain process owners or expert groups to ensure the expertise is available for all members of the department. The backbone of the department is the **Project Teams**. The teams are resourced from the staff population and exchanges can be made as the project moves through the various stages of development. Within the staff, individuals are identified as team leaders, process experts, and team representatives. The team leader and representatives make up the project team. Process experts are utilized by the teams, but can also be part of the team as a team leader or representative. Individual team leaders, process experts and representatives can contribute to several project teams simultaneously. A structure for the Analytical Department is shown in Fig. 2.1.

T. Catalano, *Essential Elements for a GMP Analytical Chemistry Department*,
DOI: 10.1007/978-1-4614-7642-9_2,

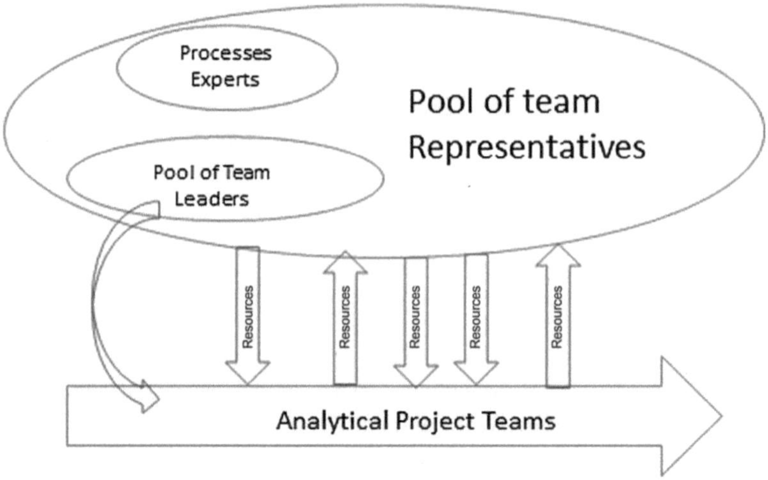

Fig. 2.1 A structure for the Analytical Department

The Process Experts interact with the project teams as shown in Fig. 2.2:

The above concepts does not negate the existence of a classical organizational chart, shown in Fig. 2.3 below, which represents the usual reporting relationships and managerial responsibilities. In this environment the reporting manager is an administrative supervisor and performance evaluations come from the team goals.

Project teams are the backbone of the organization. The department should have a specified process for the formation and management of project teams. Proper representation and functioning of project teams is essential for the success of the project to be accomplished. The Process for the development and management of Analytical Project Teams is described in Sect. 2.2.

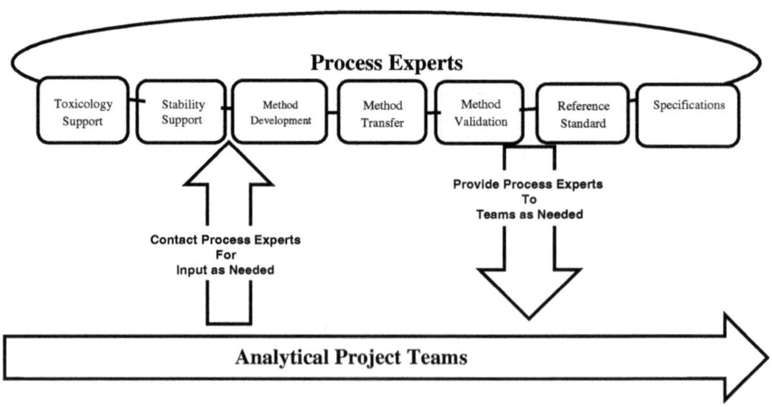

Fig. 2.2 The process experts interaction with the project teams

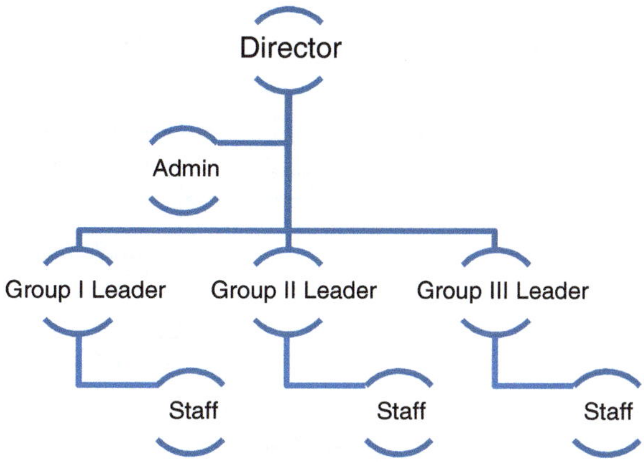

Fig. 2.3 Classical organizational chart

2.2 Project Teams

2.2.1 Project Team Development and Management

The processes for the development and management of project teams is illustrated in the following Figs. 2.4, 2.5, 2.6, 2.7, 2.8, 2.9.

The strategy document is a detailed description of the project goals and the approach the team will take to achieve these goals. The elements of the strategy document should contain some of the following items:

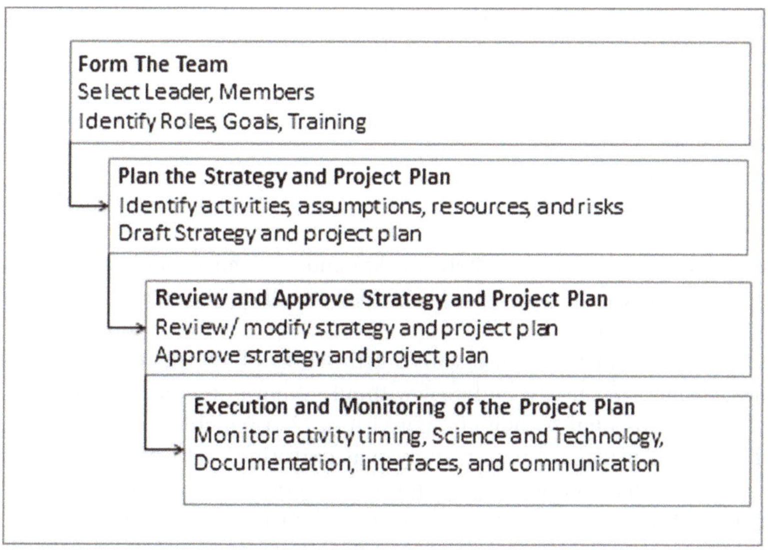

Fig. 2.4 Analytical project teams process outline

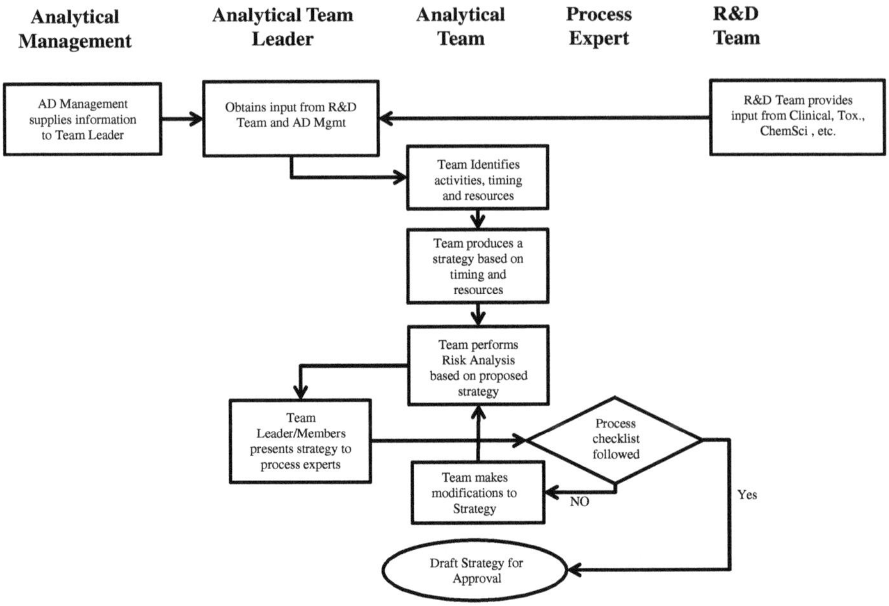

Fig. 2.5 Project strategy development process

1. Summary of the R&D project goals (IND, NDA timing etc.)
2. Chemistry manufacturing schedule
3. Potential/Expected chemistry issues (e.g. solid state, toxic imp)
4. Stability plan (chemical and dosage form)
5. Dosage Form Development (multiple dosage forms)
6. Dosage form manufacturing schedule
7. Additional studies (BA/BE, Tox., etc.)
8. Outsourcing considerations
9. Technology Transfers
10. Specifications setting schedule
11. Resource allocations.

A detailed checklist is available in Appendix I and should be utilized when developing project strategies and plans. An example of a strategy Document is shown in Appendix II.

The low risk plan is a conservative estimation of the plan elements such as, activities, timing, resources, and cost without taking into consideration any of the project goals. This plan is then compared to a plan which is required to meet the project goals and a risk evaluation analysis is performed. The plan is modified until an acceptable balance between risk and project goals is reached. The comparative risk analysis evaluation process will be described later in the chapter.

Once the project plan is approved, it is the responsibility of the team to execute the plan. The process for the execution of the plan is described in Fig. 2.9.

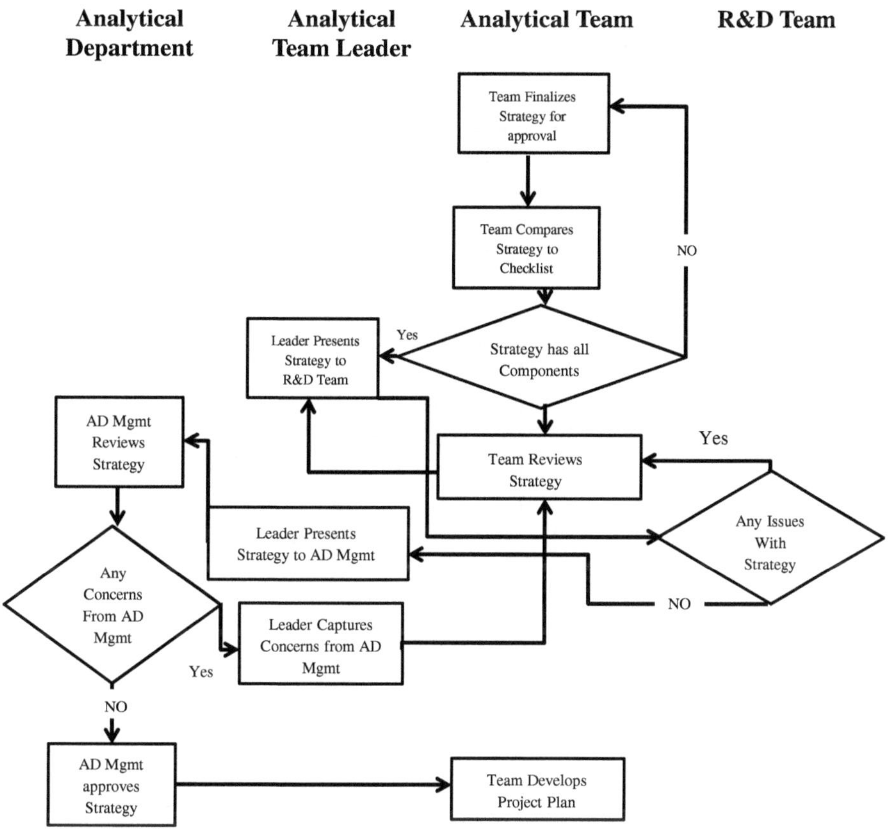

Fig. 2.6 Project strategy approval process

An important part of the execution of the plan is the monitoring of the plan. The monitoring should include items such as: key activities, target start and, actual start and completions dates, deviations from the plan, and reason for any deviations. It is important to capture whether deviations are data driven or due to lack of the team's performance. Table 2.1 illustrates the monitoring of the plan.

2.2.2 Risk Evaluation

The comparative Risk Evaluation Analysis utilizes weighted activities for each stage of development and a rate factor, which is the probability for the successful completion of the activity in the given time frame. The product of the weighted activity and the rate factor results in a Total Risk Value (**X**). The Total Risk Value (**X**) absolute value is of no significance, it is the comparative Risk Value ($\%R_c$) that is

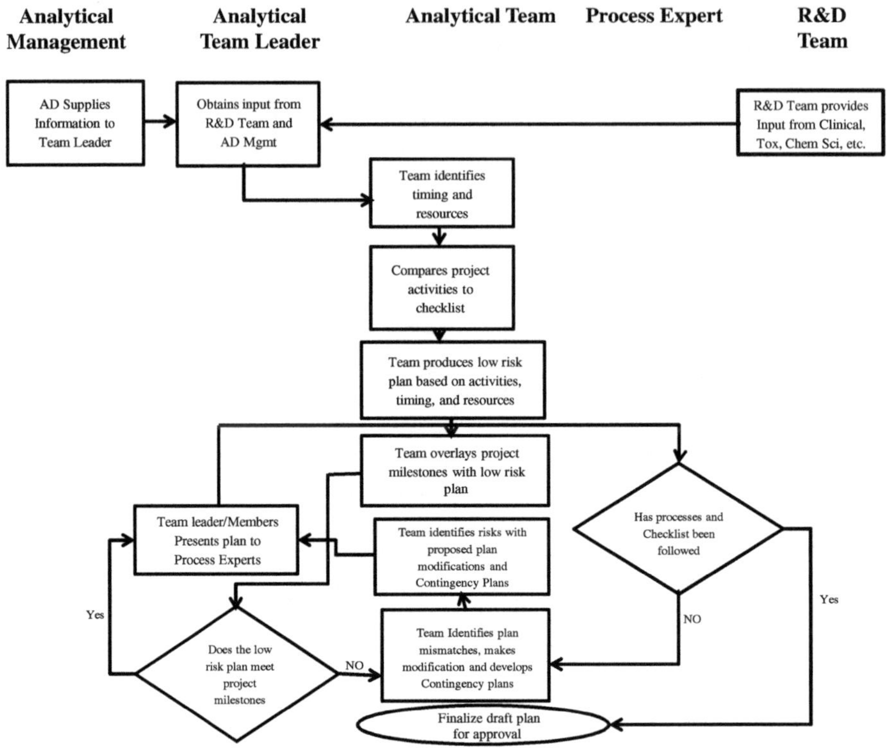

Fig. 2.7 Project plan development process

utilized to evaluate the acceptability of the new plan over the current plan. A detailed description of the Project Risk Assessment process is found in Appendix III.

2.2.3 Project Team Dynamics

There are several team dynamics which should be followed for a team to become a fully functional team [1]. They are:

1. Roles and responsibilities of the team leader and members
2. Utilization of a brainstorming process
3. Rounds of Reasoning
4. Clarifying Questions
5. Highest Level of Authority (HLA)
6. Modes of decision making
7. Use of Consensus
8. Scribe
9. Facilitation
10. Governance.

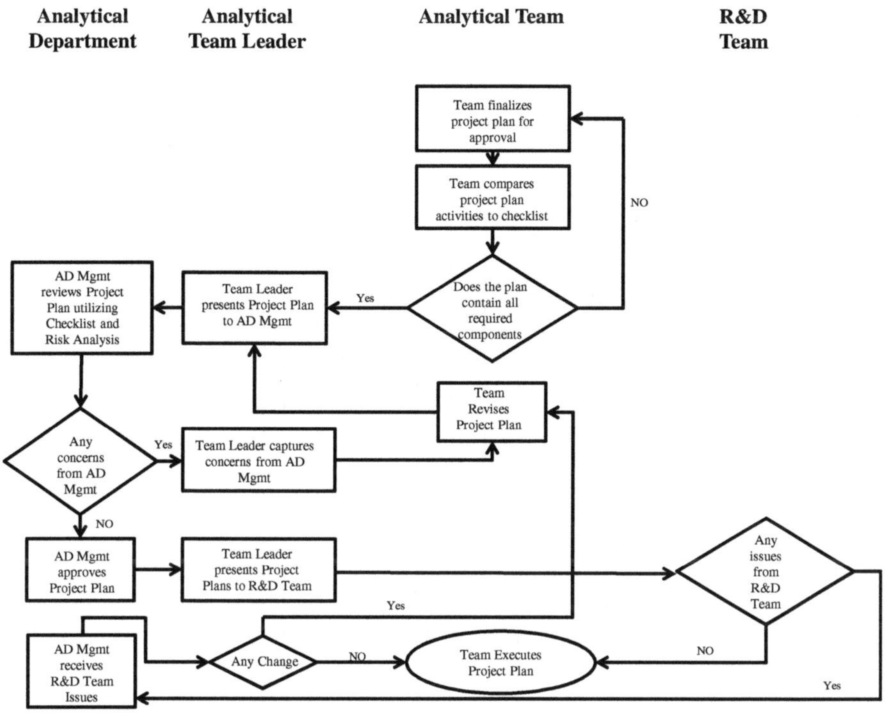

Fig. 2.8 Project plan approval process

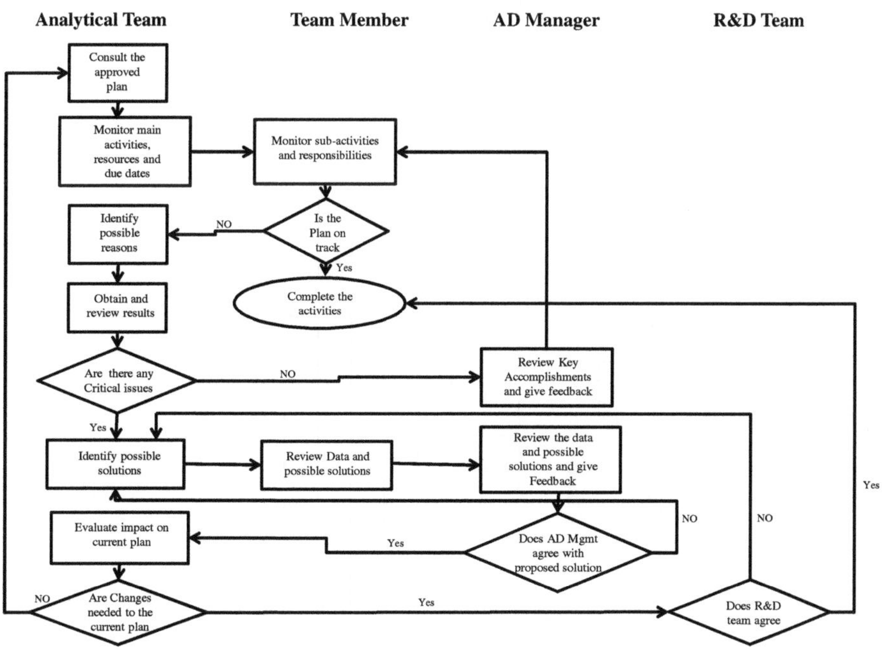

Fig. 2.9 Execution of the plan

Table 2.1 Monitor execution of the plan

No.	Key Activity	Duration	Target Start Date	Target Completion Date	Predecessor Activity	Actual Start date	Actual Completion Date	Deviation from Plan	Reason for Deviation

2.2.3.1 The Teams Leader Role is as Follows

1. Act as the leader and a communication link with management
2. Set the agenda for each meeting
3. Manages the time resource for each agenda Item
4. Responsible for the minutes for each meeting
5. Is also a member of team as a technical expert in a discipline(s)
6. Is the facilitator of the project review meetings.

2.2.3.2 The Team Member Role is as Follows

1. Is the team technical expert in a discipline(s)
2. Is also an active contributor to the team outside of their expertise
3. May be required to lead sub-teams
4. Will make presentations at the project review meetings
5. Will contribute to consensus or voting decisions
6. Is an effective communicator (written and oral).

2.2.3.3 Brainstorming Process

The Brainstorming process allows each team member to contribute by making suggestions on activities, issues, solutions etc. in an orderly fashion. The process involves going around the table allow each team member to give their suggestions. There is no counter arguments allowed from other team members, only clarifying questions could be asked. The Scribe will capture all of the suggestions on flip charts. After several rounds and there appears to be no further suggestions the team leader will end the brainstorming session and move on to a Rounds of Reasoning.

2.2.3.4 Rounds of Reasoning

Round of Reasoning is a process where each team member is allowed to challenge or support any of the suggestion captured during the brainstorming session. After several rounds a final list of suggestions is comprised and each team member is given 5 votes which they can place next to the suggestions they support. After all the votes are placed the top 5 selected suggestions are chosen and are worked by the team to obtain a final result.

2.2.3.5 Clarifying Question

These are questions which are directed towards asking for a better understanding of the issue. Clarifying questions should not be used to pass judgment or disagreement with the issue.

2.2.3.6 Highest Level of Authority

The HLA is generally part of the management team such as; Director, Associate Director, Section Head, etc.

2.2.3.7 Modes of Decision Making

There are four modes of decision:

1. A decision comes from the HLA and the team implements the decision, there is no discussion or feedback from the team.
2. A decision comes from the HLA and the feedback is requested from the team. The HLA makes the decision without addressing the feedback given by the team.
3. A decision comes from the HLA, there is discussion between the HLA and the team however, the HLA makes the final decision.
4. The HLA gives the team complete empowerment to make the decision and the HLA accepts the decision made by the team.

2.2.3.8 Use of Consensus

The consensus process requires 100 % agreement among the team members; it is not a majority rule scenario like taking a vote. Consensus can still be reached even though members of the team may not fully agree with the decision, but are will to support the team decision. Consensus is generally better than a voting process because you have the agreement that all team members are willing to support the team decision.

2.2.3.9 Scribe

The scribe role is to capture important information on to Flip charts, so the information can be placed into minutes or reports. The flip charts are generally kept as the original reference until the activity is finalized. The scribe role is usually shared among the team member based on a schedule. The scribe is still part of the team and should contribute as a team member while acting as the scribe.

2.2.3.10 Facilitation

The facilitator role is as an overseer of the team process and to help the team to stay on track by making them adhere to the process. The facilitator is not a team member and does not contribute to the agenda. The facilitator has expertise on all team dynamics and total quality management processes.

2.2.3.11 Governance [2]

Governance is the review of the project by the function management. The AD management and the analytical team make a presentation at the function project team meeting. The function management responds to the presentation with questions and concerns, AD management captures the concerns and any unanswered questions and instructs the analytical team to evaluate them against the current team project plans. If revisions to the project plans are required, the analytical team will make the necessary revisions and obtain approval from AD management. The analytical team will then schedule another meeting with the function management. If no revisions to the project plans are required the AD management will prepare a cover letter summarizing the meeting conclusions and attaches it to the finalized presentation which is submitted to the function management.

2.3 Responsibilities

The analytical department responsibilities range from supporting discovery activities, through phase's 1.2, 3 and product registration. These responsibilities are described in the timeline indicated in Fig. 2.10 [3].

2.4 Interactions [3]

The analytical department is a major contributor to a majority of functions within a Pharmaceutical Company. Its contribution is mainly in the form of data that it generates and the technology it developments and validates. Much of the conclusions made for the development of the active pharmaceutical ingredient (API) and the drug product are based on the data generated from the analytical department. The API characterization and justification for the choice of the drug formulation are based on the analytical data from methods which were demonstrated that they are appropriate for their intended use through detailed validation. Specification setting is another major responsibility of the analytical department. A detailed Specification process should be utilized, which will be discussed in a

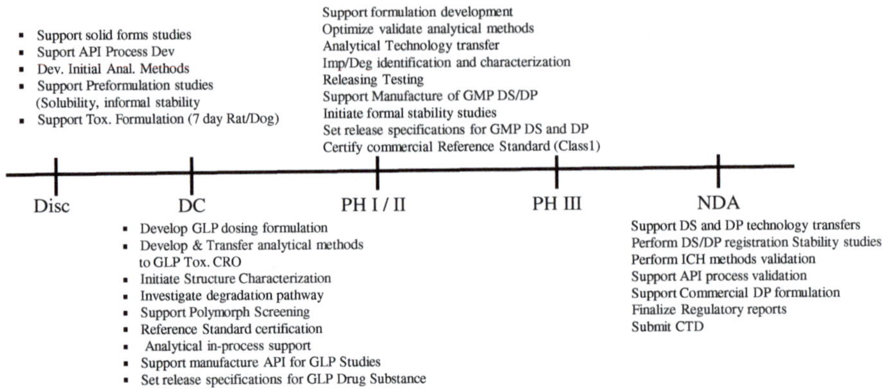

Fig. 2.10 Analytical responsibilities

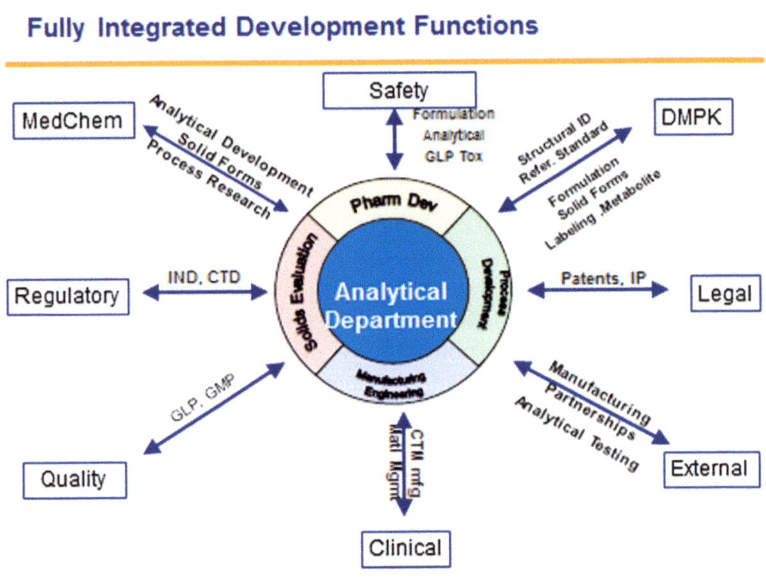

Fig. 2.11 Analytical department interactions

later chapter. Figure 2.11 is a diagram of the interactions of an analytical department within a typical pharmaceutical company.

Other Activities, within the department, which are essential to maintain staff interactions and a free flow of information, are the Staff and Project review meetings. Staff meetings are generally held bimonthly, they are generally administrative in content, an example of a typical staff meeting agenda is shown below Table 2.2.

Table 2.2 Staff meeting agenda

# Agenda Item	Type	Individual	Minutes
1 Safety			
1.1 Lab inspection	Update	NG	10
2 GMP and laboratory operation			
2.1 GMP compliance, lab operation issues, expired chemicals	Discussion	TC	15
3 Departmental operation			
3.1 Update for board visit	Discussion	ZG	5
4 Instrument and capital			
4.1 2008–2009 capital budget items	Discussion	All	10
5 Vacations and department leave			
5.1	Update	TC	5
6 General off site meetings			
6.1	Update	TC	5
6.2	Discussion	All	10

Table 2.3 Project plan

Item no.	Activity	Start date	Finish date	Total time (days)	Resources	Precursor links
1	Develop method	1/2/12	1/16/12	10	A	
2	Qualify method	1/17/12	1/1/20/12	3	A	
3	Validate method	1/21/12	1/26/12	5	A	1, 2
4	Documentation	1/20/12	2/4/12	15	0.4A, B	3
5	Method transfer	2/5/12	2/25/12	20	A, B	4
6	Release testing	2/26/12	3/6/12	10	A, 0.5B	4
7	Report of analysis	3/7/12	3/10/12	3	0.6 A	6

Project review meeting are generally held monthly and are very technical in nature. There is usually are detailed project plan and the team responds to the progress of each activity, resource issues that need to be addressed, technical difficulties, inclusion and exclusion of items within the plan, and the current risk assessment for the project success. There are several computer software programs available on the market which can support very sophisticated project plans. An example of a typical basic project plan is shown in Table 2.3.

Where each resource is identified by a letter A, B, C, etc. and the fraction before each letter represents the fraction of that resource being utilized for that activity. The precursor links represent which activities are dependent on each other and changes in one impact the other.

2.5 Operating Guidance's

Among the most basic essentials, and a requirement, in any successful GMP Analytical Chemistry Department is the establishment of Standard Operation Procedures (SOP's) and working Guidelines. This is the most efficient way to communicate best practices within the group and establish minimum standard requirements which must be met. It is a requirement and a good practice that personnel are trained on these SOP's and guidance's annually or when changes are made. In addition it is a good practice that analytical personnel are made familiar with the contents of regulatory documents in general, such as the CMC, so that it allows the staff to understand the need for the attention to detail and the stringent documentation requirements placed on them. Owing to the changed scenario in the pharmaceutical manufacturing and control procedures, coupled with the risk based approach to pharmaceutical GMPs adopted by US FDA, it essential for the analytical department SOP's and Guidelines are kept current and meet the same quality standards. The SOP's are given classifications such as, Laboratory Operations, Department Policies, and Management Control. Examples of SOP's under each classification are given below.

2.5.1 SOP's and Guidelines

2.5.1.1 Laboratory Operations

- Automated Instrument Implementation and Use
- Rounding Rules
- Decimal Place Reporting for Analytical Data
- Criteria for Identification and Qualification of Impurities
- Qualification of Chromatographic Peaks from Stability Samples as Degradation Product
- Replicate and Composite Size Determination for Dosage Form Assays
- Criteria for Identification and Qualification of Impurities
- Reporting Impurities, Including Degradation Products
- Forced Degradation Studies for Method Development
- Excipient/Raw Material Control/Acceptance Testing
- Analytical Data Review Process
- Equipment Calibration and Maintenance.

2.5.1.2 Department Policies

- Notebook/Data Handling/Creation and Use of Work Sheets
- Analysis Request/Sample Handling/Reports of analysis
- Analytical Method Development

- Analytical Method Document
- Method Validation Packages/Reports
- Performance Characteristics of Method Validation
- System Suitability for Chromatographic Methods
- Retention Sample Policy
- Laboratory Investigation of uncharacteristic analytical results
- Records Retention
- Reference Standard Certification.

2.5.1.3 Management Control

- Specifications Development
- Documents for Submission to Regulatory Agencies
- Method Transfer Process
- Analytical Support of GLP Studies
- Personnel Training and Certification program.

SOP's are written to allow for a clear understanding of the intended procedure but to an extent of detail which will allow for minor variations without being considered non-conformant. However on the other hand Guidelines are considered to be documents associated with the SOP's but the procedure is described in much more detail and allowing for little flexibility without justification. All SOP's and guidelines should be written following a standardized format (template). An example of a template is shown in Fig. 2.12.

2.5.2 Regulatory Guidance's

The history of medicinal product registration, in much of the industrialized world, has followed a similar pattern which could be described as: *Realization, Rationalization and Harmonization.*

- The Realization
  - It was important to have an independent evaluation of medicinal products before they are allowed on the market,
  - In the United States—1930s,
  - In Japan—1950s,
  - In Europe—1960s.
- Rationalization
  - Although different regulatory systems were based on the same fundamental obligations to evaluate the quality, safety and efficacy, it required duplication of many time-consuming and expensive test procedures, in order to market new products, internationally.

Company Name	Standard Operating Procedure	Document No.	Rev. 1
Standard Operating Procedures, Title		Effective Date:	

1.Purpose – define the intent of the document.

2.Scope – describe the range of systems or processes the document covers. It may also describe any exceptions.

3.Responsibilities – list the responsibilities of the individuals or departments.

3.1

4.References

4.1 List any government governance document which supports the procedure.
4.2 List any guidance or trade document which supports the procedure, such as ICH, USP/EP, etc.
4.3 List any internal document which supports the procedure.

5.Attachments

5.1 List any attachments at the end of the document
5.2 List any forms associated with the document

6.Definitions – define any acronyms and/or verbiage which may not be understood by those who train on the document.

7.Equipment and Material – list all equipment and material needed to follow the procedure. Be as specific as needed to comply.

8.Procedure – General and specific instructions are written in a chronological order.

9.Revision History- The revision history lists the Document control number, revision number, effective date, description of changes, and the initials of the initiator.

9.1 Reference to another document is **not** an acceptable description of change. The description of change needs to be as concise and complete as reasonably possible.
9.2 Do not delete any of the sections from the template. If any of the above Sections are not required, mark as N/A under section heading.

DCR #	Revision	Effective Date	Description of Change	Initiator Initials
	1		Initial Release	

Fig. 2.12 Standard operating procedure template

- Harmonization
  - Harmonization was implemented due to concerns over rising costs of health care and the need to meet the public expectation that there should be a minimum delay in having safe and efficacious new medications available to patients in need.

There are guidance's that should be followed when generating data, and writing justifications, protocols and reports. If not followed, a very strong justification must be submitted to the agency along with comparative data or logic.

- GLP's—21CFR part 58 [4]
  - Governs the conducting of nonclinical laboratory studies that support or are intended to support applications for products regulated by the Food and Drug Administration. Compliance with this regulation is intended to assure the quality and integrity of the safety data.
    There are several subparts that control all aspects of the studies performed.

    Personnel,
    Facilities (testing and operation),
    Control Articles,
    Protocols,
    Records and Reports.

- GMP's—21CFR part 210 and part 211 [5]
  - The regulations in the parts of these chapters contain the minimum current good manufacturing practice and controls used for:

    Facilities,
    Manufacture processing,
    Packaging,
    Storage,
    Distribution.

  - Drug substance and product to meet the requirements for:

    Safety
    Identity
    Strength
    Purity

  - The failure to comply with any regulations above shall render the drug substance and/or drug product to be non-compliant as well as the person who is responsible for the failure to comply; both could be subject to regulatory action.

- International Conference on Harmonization (ICH) [6]
  They consist of Guidelines Q1 to Q10. The ICH Guidelines are complimentary to The GLP's and GMP's; they describe in more detail the activities to be performed and the criteria to be achieved such as
  - Stability (Thermal and Photo)

    Bracketing and Matrixing designs for stability testing
    Evaluation of stability data
    Stability data for Climate Zones III and IV

- Validation of Analytical procedures

    Attributes
    Criteria

- Specifications

    Critical Attributes
    Justification

- Impurities

    Control new drug substance and product impurities
    Definitions (Qualified, Specified, Unspecified
    Residual solvents

Class 1, Class 2, Class 3

- Good Manufacturing Practice

    Quality management
    Personnel
    Building and facilities
    Material Management
    Laboratory Controls
    Process Validation Program
    Cleaning Validation Program
    Microbiological attributes

- Pharmaceutical Development

    Components of the Drug Product
    Manufacturing Process Development
    Container Closure System
    Compatibility

- Quality Risk Management

    Scientifically Based
    Based on level of risk

The regulatory guidance that the **GMP Analytical Chemistry Department** plays a significant role is the submission of the **Chemistry, Manufacturing and Controls (CMC)**. The Chemistry, Manufacturing and Controls (CMC) is a compilation of information, data, justifications and reports required to adequately characterize drug substance and drug product for acceptance by global regulatory agencies for approval to be commercially available to the public The generation of information, data, justifications and reports can begin at the discovery stage and continue throughout development process.

The following Fig. 2.13 is an example of the contents for a typical CMC submission.

Table of Contents

3.2.S Drug Substance

- ♦ 3.2.S.1 General Information
  - 3.2.S.1.1 Nomenclature
  - 3.2.S.1.2 Structure
  - 3.2.S.1.3 General Properties
- ♦ 3.2.S.2 Manufacture
- ♦ 3.2.S.3 Characterization
- ♦ 3.2.S.4 Control of Drug Substance
- ♦ 3.2.S.5 Reference Standards or Materials
- ♦ 3.2.S.6 Container Closure System
- ♦ 3.2.S.7 Stability

3.2.P Drug Product

- ♦ 3.2.P.1 Description and Composition of the Drug Product
- ♦ 3.2.P.2 Pharmaceutical Development
- ♦ 3.2.P.3 Manufacture
- ♦ 3.2.P.4 Control of Excipients
- ♦ 3.2.P.5 Control of Drug Product
- ♦ 3.2.P.6 Reference Standards or Materials
- ♦ 3.2.P.7 Container Closure System
- ♦ 3.2.P.8 Stability
- ♦ 3.2.A Appendices

Fig. 2.13 Contents for CMC

The Analytical Department supplies data and technology to the CMC from three major areas.

- Process Chemistry
- Solid Forms
- Pharmaceutics and Formulation Development.

The support to Process Chemistry involves preforming release testing, impurity identification and characterization, stability and reference standard certification. The support also involves in-process monitoring by the use of Process Analytical

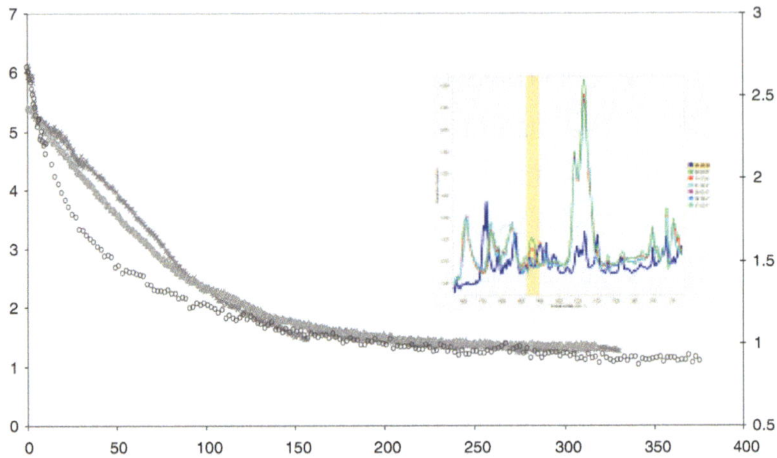

Fig. 2.14 Reaction monitoring

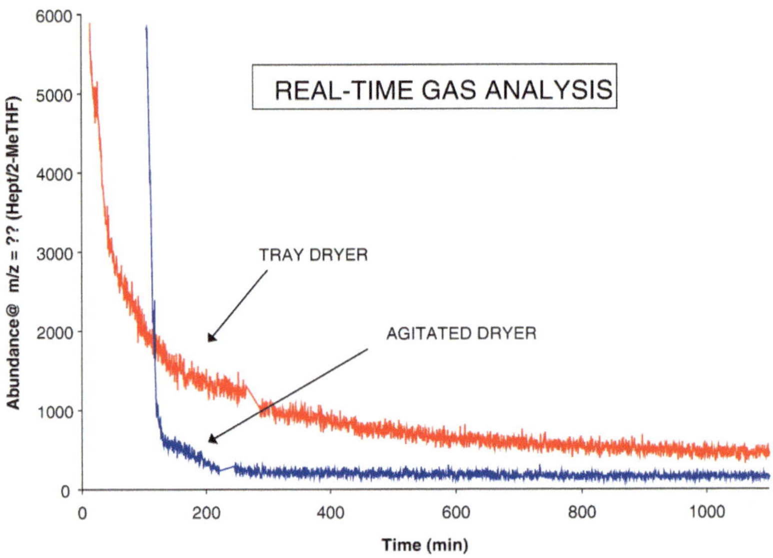

Fig. 2.15 Drying profiles

Technology (PAT). PAT utilizes in line instrumentation to monitor the process in real time and determine the critical process parameters. Examples of in-process support utilizing PAT technology is shown in Figs. 2.14, 2.15, 2.16, 2.17.

The support to solid forms involves the following activities:

1. Crystal and Polymorph Screening
2. Solid State characterization

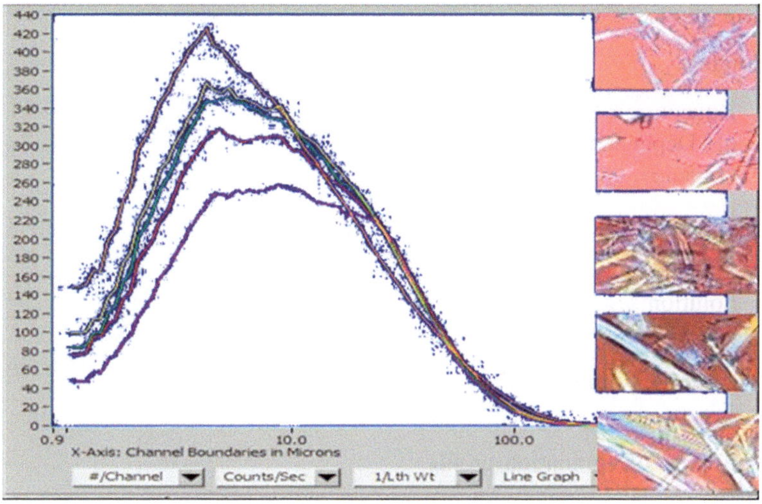

Fig. 2.16 Particle size control

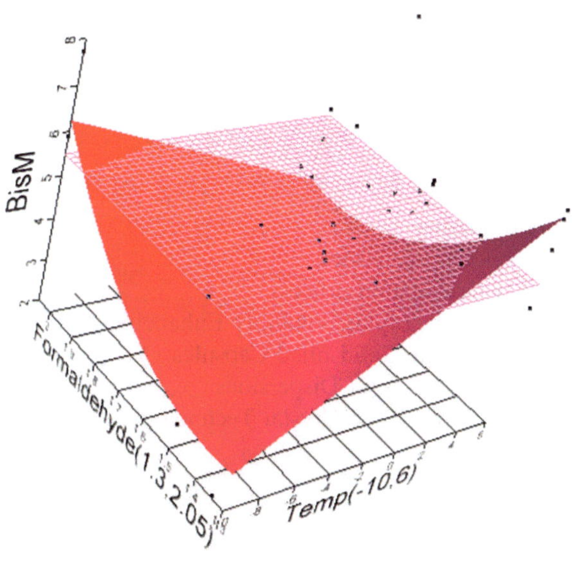

Fig. 2.17 Determination of critical process parameters

3. New Salt selection
4. Generation of intellectual Property.

The support to Pharmaceutics and Formulation Development is described below:

- Pharmaceutics
  - Solubility

    Aqueous
    pH Dependence
    Organic

  - Stability

    Solid State
    Solution
    pH Aqueous

  - Intrinsic Dissolution
  - Purity

    Impurity/Degradation profile

  - Physical characterization

    Bulk solid properties

- Formulation Development
  - Evaluate Chemical and Aqueous Stability
  - Evaluate Excipient Compatibility
  - Consider Dosage Forms

    Powder in a Capsule
    Powder in a Bottle
    Blended Formulation
    Evaluate Form Change Issues
    Consider Wet Granulation process
    Evaluate Stability in Lipids/Solubility Enhancing Agents
    Consider Liquid/Semi Solid filled Capsules
    Consider Amorphous Solid Dispersions
    Consider Micronized Nano-Crystal Technology.

References

1. Coates and Freeman Inc., (1998) Total quality management consulting. G.D. Searle Inc., Chicago
2. Joiner B (1997) 4th generation management. Rosemount Horizon
3. Pharmaceutical Manufactures Association (1998) Pharmaceutical management development seminar, Columbia University, Arden House, Harriman, NY
4. GLP's, 21CFR part 58
5. GMP's, 21CFR part 210 and part 211
6. ICH Guidelines, Q1 to Q10

Chapter 3
Processes

Abstract The Processes that are implemented to manage the projects are presented within this chapter. The processes are divided into three categories, Safety, Technology, and Department. Each of the processes included are described in enough detail to be directly implement. Several of the process flow charts are shown along with examples of implementing the processes. The technical processes described are HPLC and Dissolution method development, Method Validation and Technology Transfer. The department processes described are Specification setting, Stability Management, Reference Standard Certification, and Training. The implementation of these processes or other processes developed following the model described will enhance the performance of the project teams and the success of the project.

3.1 Safety Process

Safety is the #1 priority in an Analytical Department. Safety violations are seriously investigated and severely reprimanded. The safety process begins with the identification of two staff members as safety monitors. These monitors have the following responsibilities:

1. Members of the company/function safety committee
2. Relate all safety information to the department
3. Train/mentor department members on all safety related activities
4. Conduct monthly safety audits within the department
5. Report all safety violations
6. Complete the department monthly safety report document
7. Manage a department quarterly safety review meeting.

T. Catalano, *Essential Elements for a GMP Analytical Chemistry Department*,
DOI: 10.1007/978-1-4614-7642-9_3,

Monitored safety concerns

- Storage of all chemicals
  - Acid and Bases—should be stored in separate storage bins marked ACID, BASE
  - Solvents—should be stored in metal cabinets, if they need to be refrigerated store in explosion proof refrigerator
  - Prepared material—should be stored under the fume -hood in containers to avoid spilling

    Buffers
    Mobile Phases
    Reagents

  - Solid chemical storage room—should be a well-ventilated and climate controlled room
  - Chemical waste disposal
- Transportation of chemicals
  - Using a cart—The cart should have raised edges on all sides. Two people should accompany the cart during transportation.
  - When transporting between floors, always use the freight elevator. Do Not Accompany the cart on the elevator, include a sign on the cart stating solvents in transport "Do Not Enter".
  - Transporting chemicals by hand—Use a plastic carrier and only one carrier at a time.
- Personal protective equipment
  - Laboratory Coat—Always wear lab coat when in the laboratory and when transporting chemicals
  - Do not wear lab coat in common areas such conference rooms, cafeteria, and office areas
  - Eye glasses—Wear safety glasses with side shields while in the laboratory
- Labeling
  - Vender supplied materials

    Ensure that identity and storage conditions are legible. Label should include the following information:

    Date opened
    Analyst initials
    Expiry date, generally 2 years from date opened (but cannot exceed the manufactures expiry date)

  - In-house prepared materials label should include the following:

    Notebook reference
    Preparers initials

Identity
Date prepared
Concentration
Expiration date (not to exceed one year)
Storage conditions

- Laboratory containers of solvents (e.g., squirt bottles) for temporary use must be labeled with identity and concentration of mixtures

  Solvent containers are re-filled monthly
  Water containers are re-filled weekly

- In-house prepared samples and standard solutions expiry date must be based on validation data. If validation not yet performed then a maximum of 2 day expiry may be assigned.
- Vials or flasks too small for conventional labels, should be grouped and a label with conventional information made for the group. However, each individual vial or flask should have a label with an identifying mark

• Investigational materials label should include the following:

  - Notebook reference
  - Preparers initials
  - Identity
  - Date prepared
  - Concentration
  - Expiration date (not to exceed one year)
  - Storage conditions

• Glassware policy

  - Rinsing contaminated glassware

    Triple rinse with appropriate solvent
    If the rinse solvent other than water, perform an additional final rinse with water

  - Non contaminated broken or non-broken glassware should be placed in a container labeled a non-contaminated glassware
  - Contaminated glassware which has not been rinsed should be placed in a bucket and labeled "contaminated glassware for disposal"

• Solvent handling

  - Solvent transfers should be done under the fume hood
  - Wear appropriate gloves when handling solvents

• Sample weighing

  - Weigh all chemical utilizing a local exhaust (balance enclosure) or in the fume hood.

- Housekeeping
  - Good housekeeping is extremely important, it is the 1st impression made by an auditor when he walks into the laboratory.
  - All samples and chemicals should be stored appropriately at the end of the day or when not being utilized. Chemical and samples should not be left on the bench top.
  - Glass bottles and other heavy items should not be stored on shelves above eye level.
  - All drawers and cabinets should be labeled with their contents.
  - Floors should have tape signifying laboratory areas from non-laboratory areas.
  - All laboratory floors and bench tops should dusted and cleaned at the end of each day.

3.2 Technology Processes

3.2.1 Systematic Approach to HPLC Method Development [1, 2]

HPLC is one of the most utilized analytical methodology in a pharmaceutical analytical laboratory. The data generated from the HPLC methods impacts the products from discovery to commercialization. It is the goal for every Analytical Department to develop HPLC methods that are acceptable for their intended use as rapidly as possible. The use of a systematic approach to HPLC method development, as described below, has many advantages over the trial and error approach. Experience in the development of HPLC methods is also of great importance and when coupled with a systematic approach it becomes a very powerful and valuable tool. The use of a systematic approach has the following advantages:

- Development of chromatographic data base
- Increase in the quality of data at early stages of development
- Enhance the development of future methods
- Rationale for the methods developed
- Good training tool
- Standardize the approach
- Addresses a minimum acceptable criteria for all methods.

Before starting method development the following items should be considered:

- What is the intended use for the method
- Determine Time available for development
- Gather any existing information
- Gather samples required for method development
- Establish method criteria.

Examples of existing information which should be obtained if available are as follows:

- Chemical structure
- Physiochemical properties
- Synthesis
- Dosage form
- Literature/References
- Toxicology dosing strategy
- Current methods
- Related methods
- Sample analysis requirements.

Acquiring the appropriate samples as early as possible in the development process is crucial to the success of producing an acceptable method. Examples of samples which may be appropriate are as follows:

- Reference standards
- Authentic material
- Chemical lab lots
- Starting material
- Intermediates
- Potential impurities
- Mother liquors
- Formulated material
- Placebo
- Dosage form
- Delivery device
- Capsule shell
- Transdermal pad components including pad, adhesive
- Excipients
- Degraded chemical
- Degraded dosage forms
- Degraded placebo
- Degraded solutions
- Degraded vehicle
- Isolated degradation products.

Methods may require revision at different stages of development. In early development there is more concern with the method selectivity as compared to accuracy and analysis time. Examples of method criteria for early development are as follows:

- HPLC gradient method (preferably ACN)
- Parent separated from impurities/degradation products
- Minimum quantifiable limit is approximately 0.5 %
- Minimum detectable limit is approximately 0.1 %

- Evaluate mass balance
- Non-validated method.

At later stages of development the criteria changes. The suggested criteria for API at pre-clinical and clinical stages are as follows:

- HPLC isocratic method is preferred
- Parent is separated from impurities/degradation products
- Impurities/degradation products are separated from each other
- Accuracy is within the range of 97–103 %
- Minimum quantifiable limit $\leq$0.05 %
- Minimum detectable limit $\leq$0.01 %
- Reasonable mass balance >95 %
- Analysis time <60 min
- Validated method.

For Drug product the suggested criteria are:

- HPLC isocratic method is strongly suggested
- Parent is separated from degradation products
- Degradation products are separated from each other
- Accuracy within the range 95–105 %
- Minimum quantifiable limit $\leq$0.1 %
- Minimum detectable limit $\leq$0.05 %
- Reasonable mass balance >95 %
- Reasonable analysis time <45 min
- Method validated.

The mobile phase is a critical component of an HPLC method and there are many properties that should considered such as:

- Solvent UV cutoff
- pH
- Buffer UV cutoff
- Ionic Strength
- Ion-pairing Reagent
- Background Absorbance/Fluor
- Viscosity
- Column Compatibility
- Compatibility with MS.

Isocratic Method Development Process [3, 4]

- Obtain appropriate Samples
- Select Columns for evaluation (e.g., C18, Phenyl, CN)
- Setup an Initial ACN/Buffer Gradient
  - 5–95 %, 20 min, 10 min hold

- Observe if any peaks have retention time at the hold time
  - If any peaks observed during hold time, revise gradient for each column evaluated until no peak retention times are found in the hold time
- Choose Column that gives greatest number of peaks and best selectivity
- Determine whether a Isocratic method feasible for development
  - Range of all peak retention times in gradient run must be ≤40 % of the total gradient time
- Calculate the isocratic solvent strength for the main peak in the ACN gradient
- Run the isocratic condition and determine the k′ for 1st and last peak
- Adjust the Isocratic solvent strength to give a k′ of approximately 2–20 for 1st and last peak respectively
- Calculate equivalent solvent strength for MeOH and THF using the Solvent Strength Conversion Chart in Table 3.1
- Construct a ten Experiment Solvent Mixture Design Triangle based on the equivalent solvent strengths at the corners obtained from the Solvent Strength Chart (See Table 3.1)
- The sides of the triangle consist of 66 and 33 % of the solvent corners making up that side. The middle is 33 % of each corner of the triangle
- Run each experiment in mixture design triangle (Figs. 3.1, 3.2).

Table 3.1 Solvent strength conversion chart (Reverse phase)

MeOH (%)	ACN (%)	THF (%)
5	5	3
10	10	7
15	15	10
20	19	14
25	24	17
30	29	21
35	34	24
40	39	27
45	44	31
50	48	34
55	53	38
60	58	41
65	63	44
70	68	48
75	73	51
80	77	55
85	83	58
90	87	61
95	92	65
100	97	68

Exp #1
10.0% MeOH
0.0% ACN
0.0% THF
90.0% Buffer

Exp #2
6.7% MeOH
2.3% ACN
0.0% THF
91.0% Buffer

Exp #3
6.7% MeOH
0.0% ACN
1.7% THF
91.7% Buffer

Exp #4
3.3% MeOH
4.7% ACN
0.0% THF
92.0% Buffer

Exp #5
3.3% MeOH
2.3% ACN
1.7% THF
92.7% Buffer

Exp #6
3.3% MeOH
0.0% ACN
3.3% THF
93.3% Buffer

Exp #7
0.0% MeOH
7.0% ACN
0.0% THF
93.0% Buffer

Exp #8
0.0% MeOH
4.7% ACN
1.7% THF
93.7% Buffer

Exp #9
0.0% MeOH
2.3% ACN
3.3% THF
94.3% Buffer

Exp #10
0.0% MeOH
0.0% ACN
5.0% THF
95.0% Buffer

Fig. 3.1 Ten experiment solvent triangle

Data interpretation of the ten experiments is as follows:

The lower right portion of the triangle shows the greatest number of peaks and best selectivity (experiment # 6, 8, 9, 10).

It also appears that the combinations of ACN and THF give better selectivity than the combinations MeOH and THF. Therefore minor adjustments between the content of ACN and THF (experiment # 9 and 10) resulted in the final mobile phase condition of

ACN/THF/Buffer (2 %/4 %/94 %)

These experiments were performed automated over a weekend and the interpretation with confirmatory analyses over next 3 days.

The Chromatograms are shown in Fig. 3.3.

Gradient HPLC Method Development Process [4–6]

- Obtain appropriate Samples
- Select Columns for evaluation (e.g., C18, Phenyl, CN)
- Setup an Initial ACN/Buffer Gradient
  - 5–95 %, 20 min, 10 min hold
- Observe if any peaks have retention time at the hold time
- If any peaks observed during hold time, revise gradient for each column evaluated until no peak retention times are found in the hold time
- Run a Methanol Gradient of equivalent solvent strength to the ACN gradient

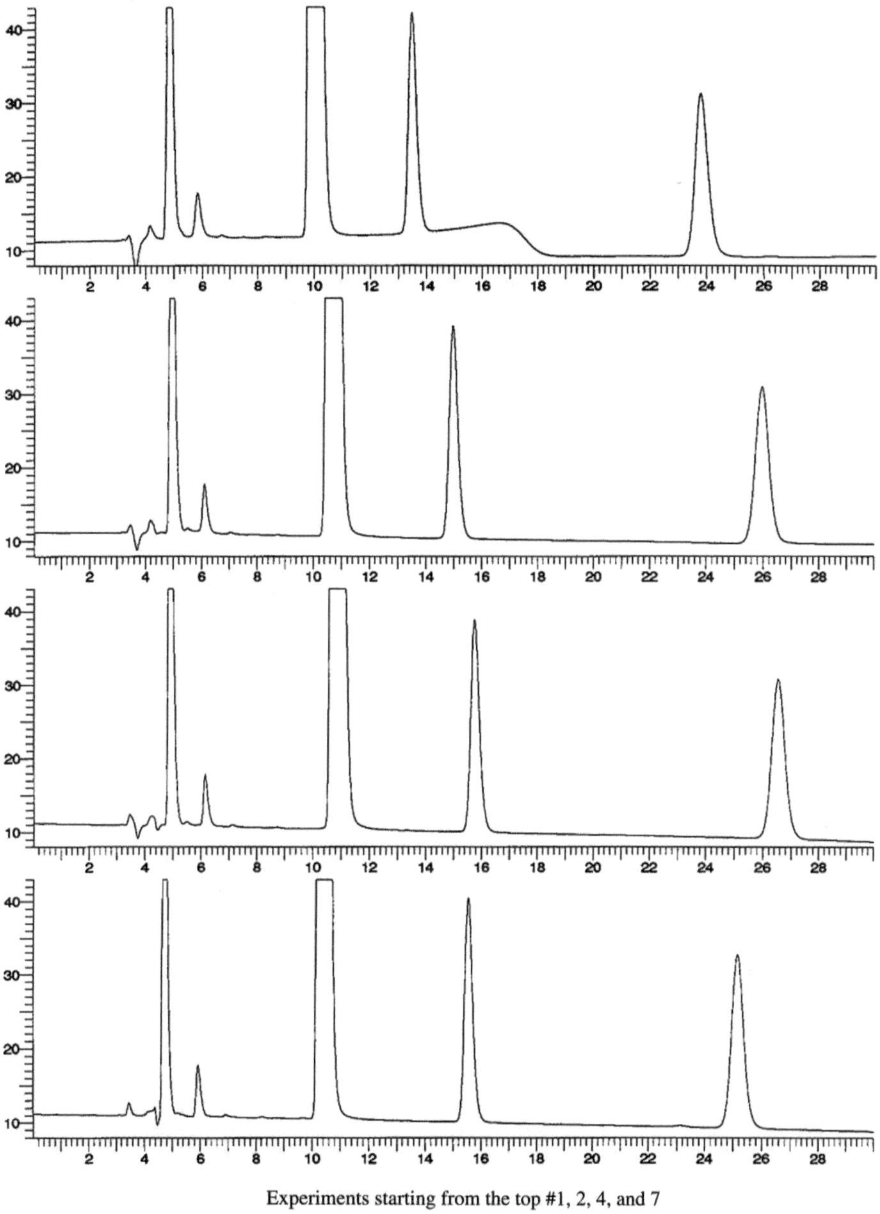

Fig. 3.2 The results of the ten experiments

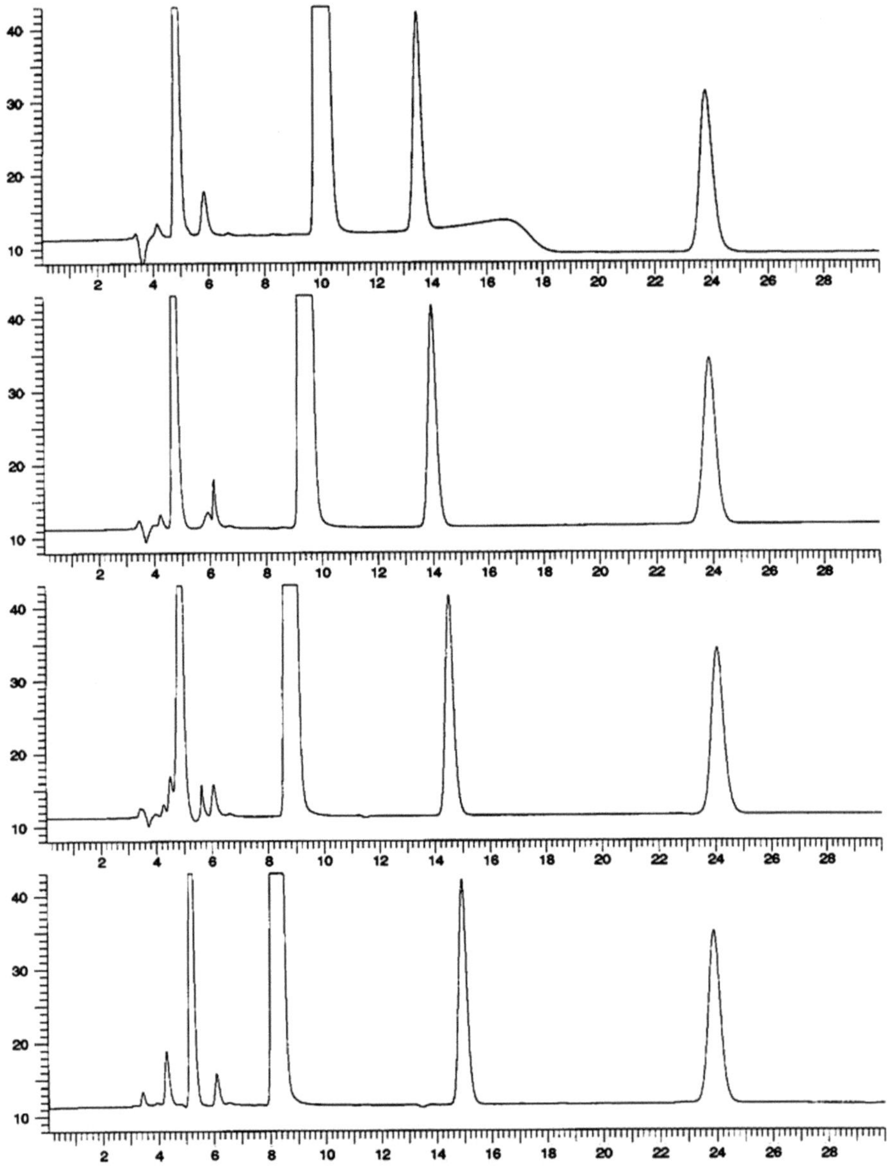

Experiments starting from the top1, 3, 6, and 10

Fig. 3.2 (continued)

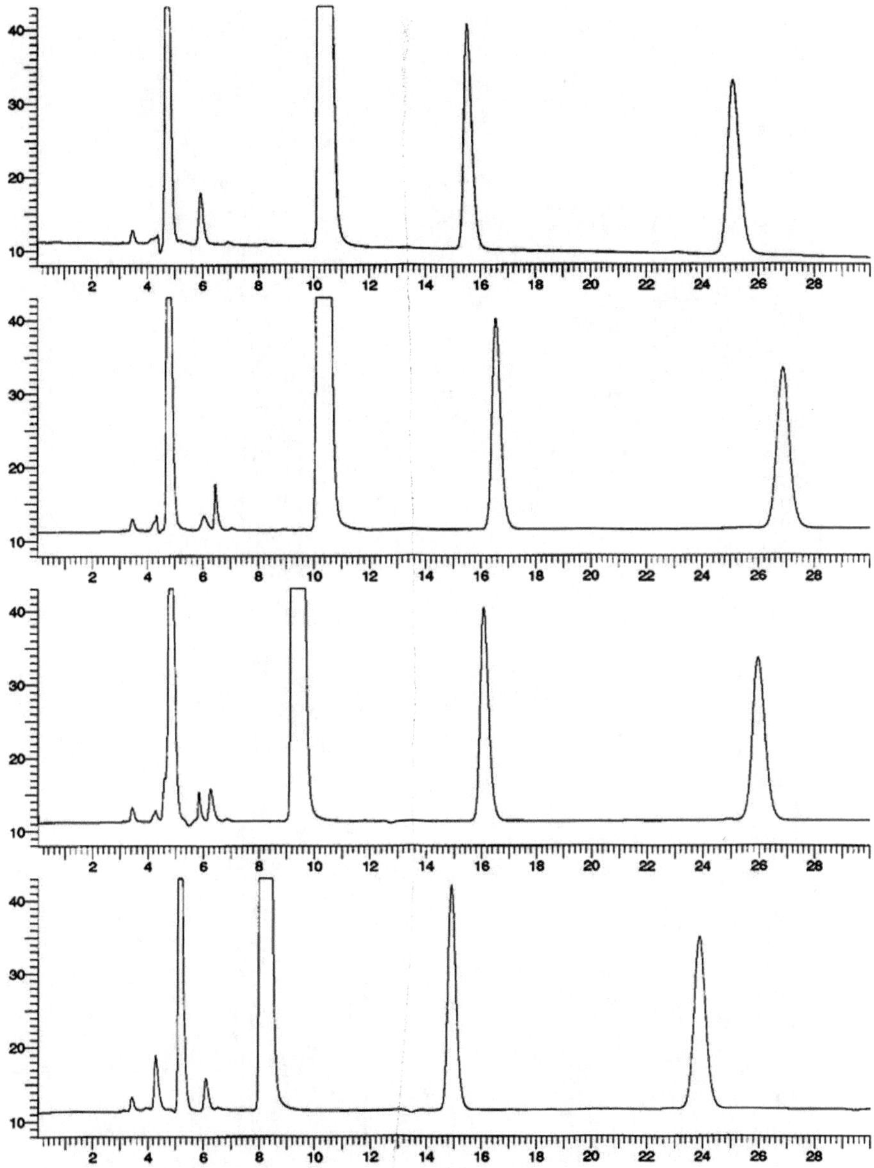

Experiments starting from the top #7, 8, 9, 10, and 5

Fig. 3.2 (continued)

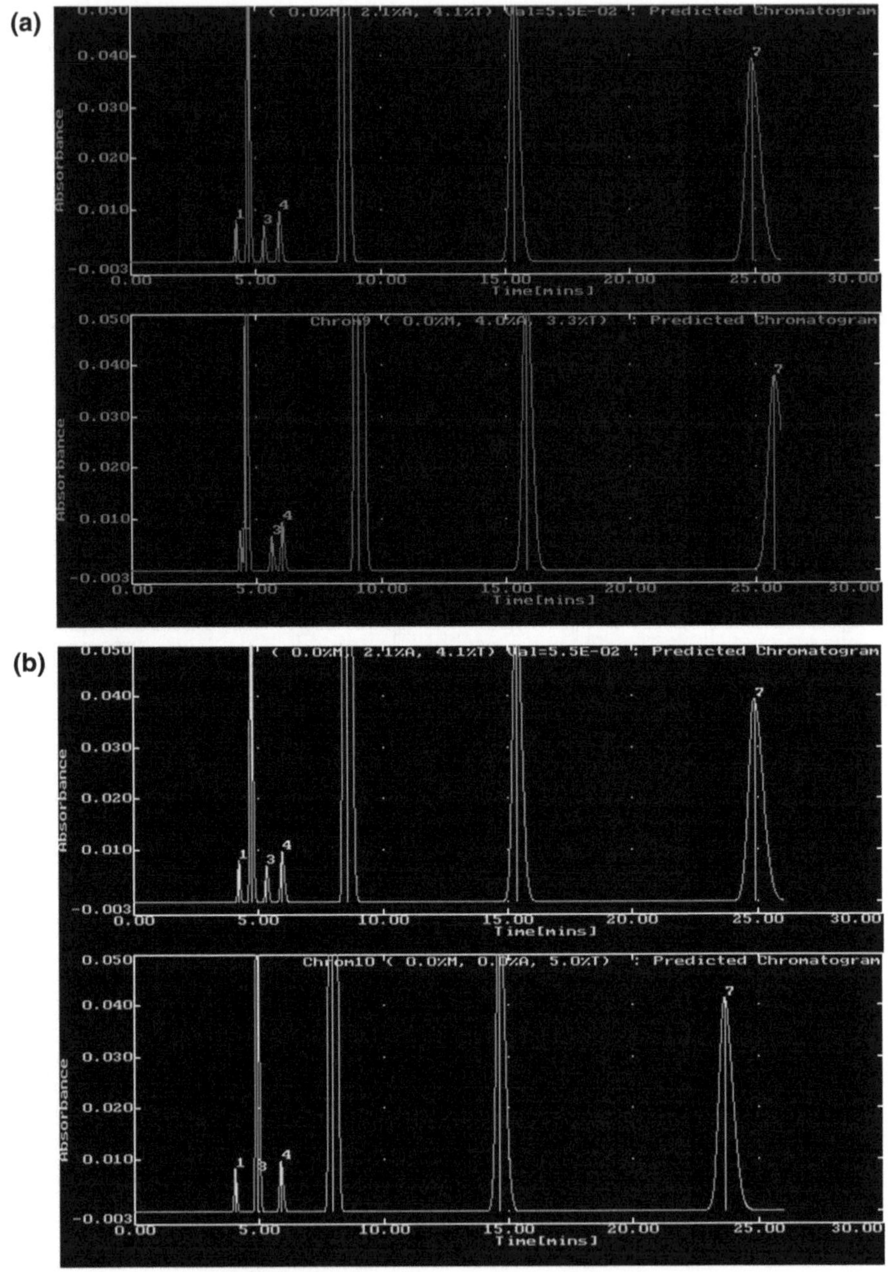

Fig. 3.3 Chromatograms (**a**) The top Chromatogram is the final condition and the bottom chromatogram is experiment # 9. (**b**) The top chromatogram is the final conditions and the bottom chromatogram is experiment # 10

- Choose Column that gives the greatest number of peaks and best selectivity
- Choose which solvent gradient (ACN or MeOH) that displays greatest number of peaks and best selectivity
- Adjust slope of the chosen gradient by changing initial and final solvent percentages so that selectivity is maintained with a minimum gradient time
- If the separation of critical peak pairs is not achieved attempt to include isocratic hold times within the gradient
- If adequate separation is still not achieved investigate mix solvents gradient (e.g., ACN/MeOH) holding the solvent strength of the mixture equivalent to the solvent strength chosen for single solvent gradient.

Examples are shown in Fig. 3.4:

Figure 3.5 is an example of a Gradient method developed using this systematic method development approach.

Sample preparation is another essential component of the final HPLC method, especially for the drug product analysis. The follow scheme shown in Fig. 3.6 is a systematic approach to sample preparation.

Once the method is finalized, a method qualification is performed to evaluate if the method is acceptable for its intended use and is validatable. The following characteristics should be evaluated:

- Analyte Stability in Mobile Phase
- Linear Range
- Accuracy/Recovery
- Sensitivity
- Initial System Suitability Parameters
- Support Specifications (precision).

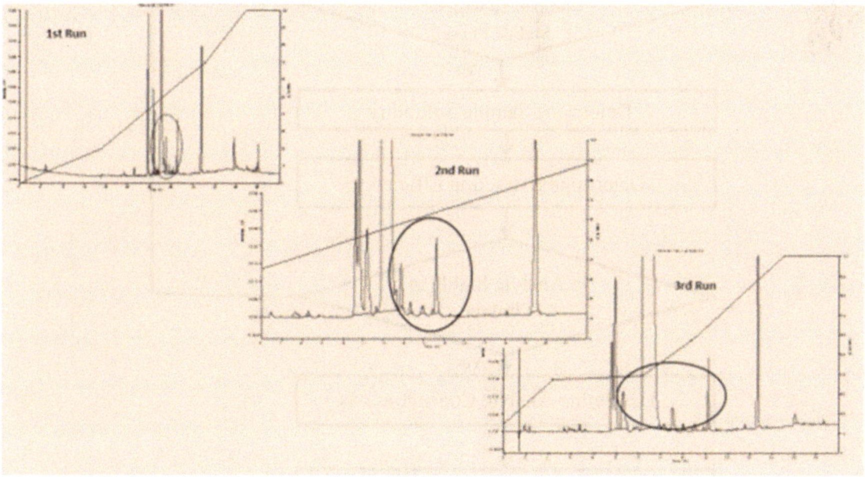

Fig. 3.4 Gradient method development. * Courtesy of ChromSword

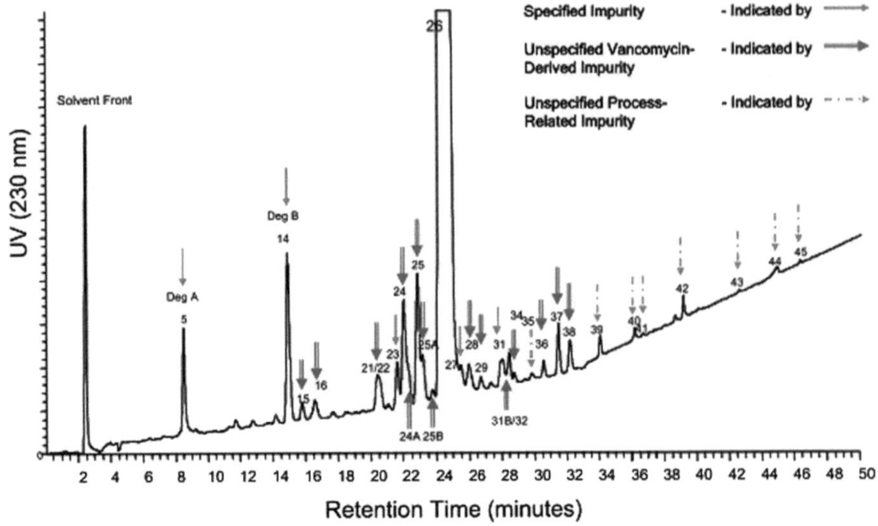

Fig. 3.5 Developed gradient method

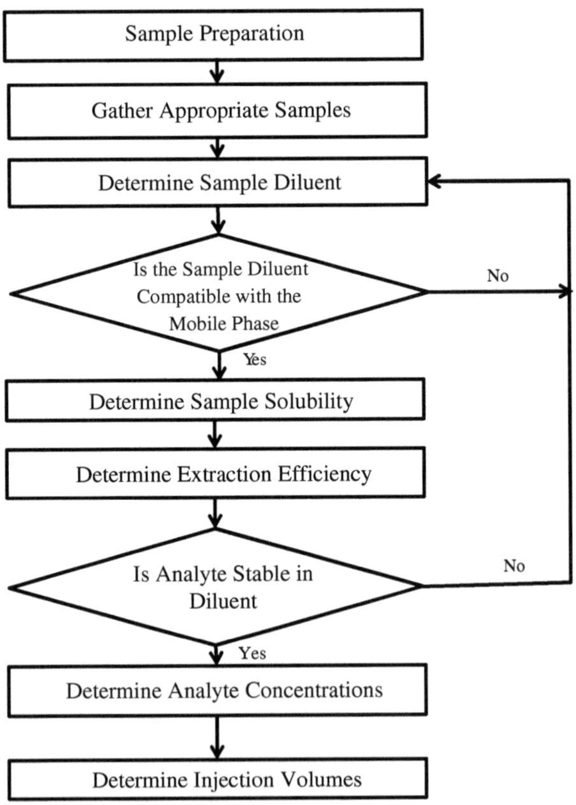

Fig. 3.6 Sample preparation process

System suitability is a regulatory requirement which is intended to show that your method is in control each time it is implemented. There are several considerations for system suitability, depending on the intended use of the method. Generally, the parameters chosen are those which are most important to control, thus obtaining acceptable results. Some of the parameters to consider are listed below:

- Resolution
- Theoretical Plates
- Tailing Factor
- Peak Asymmetry
- Standard Checks
- Limits of Detection
- Limits of Quantitation
- System Noise
- System Base Line Drift
- Injection Precision
- Retention Time Reproducibility
- Recovery Criteria
- Dwell Volume.

Once the method is qualified, the method is ready for validation.

3.2.2 Dissolution Method Development

Dissolution is one test in the pharmaceutical testing arsenal that can help determine both product quality and performance.

- As a Quality Control test for Drug Product and Drug substance one can evaluate properties such as:
  - Solubility
  - Intrinsic Dissolution
  - Particle Size
  - Excipient Differences
  - Manufacturing Differences
- Product performance (in vivo/in vitro correlation) can be evaluated based on the properties such as:
  - Permeability
  - Bioavailability
  - Partition Coefficients

- Dissolution and Disintegration are tests which can be used for the products quality control. However dissolution is the method of choice unless the following criteria are met:
  - Highly soluble compound
  - Rapid dissolution—≥85 % in 15 min
  - Must establish relationship between dissolution and disintegration by acquiring sufficient data during development
- Sink conditions
  - Generally, when developing a dissolution procedure, one goal is to have sink conditions, which is defined as the volume of the medium at least three times that required in order to form a saturated solution of drug substance. When sink conditions are present, it is more likely that dissolution results will reflect the properties of the dosage form. Lack of sink conditions may result in unwanted solution reactions such as, particle size interactions, concentration gradients, precipitation, and non-equilibrium.
  - A medium that fails to provide sink conditions may be acceptable (this is an unusual situation) if it is shown to be more discriminating or otherwise appropriately justified.
  - Assess the solubility of drug substance in the presence of the correct excipient ratio, utilizing the selected dissolution media. The measured solubility must be greater than three times higher than the highest dose strength divided by the selected vessel volume.
- Media Selection for immediate release dosage forms
  - Conduct a single dissolution experiment for each of the following three media, utilizing dosage forms which are stressed, different particle sizes, different excipients, or different manufacturing processes to determine discriminating power:

    0.1 N HCl
    100 mM phosphate, pH 4.5
    100 mM phosphate, pH 6.8

  - Purified water is not recommended as a suitable dissolution media for several reasons. First, the quality of water can vary depending on the source of the water, and the pH value of the water is not controlled. Second, the pH value can vary from day to day and can also change during the run, depending on the active substance and excipients.
  - In some instances, simulated gastric or intestinal fluid (with or without enzymes) and surfactants (with or without acids or buffers) such as polysorbate 80, sodium lauryl sulfate and bile salts may be selected as media for specialized applications. The use of enzymes in the dissolution medium is

permitted in accordance with USP <711> when dissolution failures occur as a result of cross-linking with gelatin capsules or gelatin coated products.

- Utilize a paddle speed of 50 rpm, a bath temperature of 37 °C and collect samples at 5, 15, 30, 45 and 60 min. At the 60 min time point raise the paddle speed to 250 rpm, this is known as the infinity time point and is intended to ensure complete dissolution. If dissolution is not complete repeat the experiment with paddle speed of 75 and 100 rpm. Paddle speeds of greater then 100 rpm is not recommended since it is not consistent with the physiological condition.
- Choose the media and paddle speed which gives complete dissolution within 60 min and the best discrimination.
- For high solubility compounds (class I and III), the procedures outlined above are directly applicable. For low solubility compounds (class II and IV), aqueous solutions may require a surfactant (e.g., sodium lauryl sulfate, polysorbate or laryldimethylamine oxide) to enhance drug solubility.
- Discriminating properties of the dissolution test method
  - The dissolution method can be useful in detecting deviations or errors from the manufacturing processes. To ascertain the methods discriminating properties, the following attributes should be provided by the method:

    Differences in the solid-form properties of the drug substance (hydrates, polymorphs, particle size, surface area, bulk density, etc.)
    Differences in the properties or ratios of excipients
    Differences in the properties of capsule shells

  - For dissolution method development at an early development stage, where none of the materials are yet available. Challenge the method by thermally stressing the solid dosage form in order to simulate a "mis-manufactured" dose unit.

- **Observations**
  - Visual observations and recordings of product dissolution behavior are very useful because dissolution patterns can be indicative of variables in the formulation or manufacturing process. To accomplish visual observation, proper lighting (with appropriate consideration of photo-degradation) of the vessel contents and clear visibility in the bath are essential.
  - Documenting observations by drawing sketches and taking photographs or videos can be instructive and helpful for those who are not able to observe the real time dissolution test. Observations are especially useful during method development and formulation optimization. Examples of typical observations include, but are not limited to, the following:

    Uneven distribution of particles throughout the vessel. This can occur when particles cling to the sides of the vessel, when there is coning or mounding directly under the paddles,
    when particles float at the surface of the medium,

when film-coated tablets stick to the vessel, and/or when off-center mounds are formed.
Air bubbles are formed on the inside of the vessel or on the apparatus or dosage unit. Sheen on the apparatus is also a sign of air bubbles. This observation would typically be made when assessing the need to deaerate the medium.
Dancing or spinning of the dosage unit, or the dosage unit being hit by the paddle.
Adhesion of particles to the paddle, which may be observed prior to or upon removal of the stirring device at the end of the run.
Pellicles or analogous formations, such as transparent sacs or rubbery, swollen masses surrounding the capsule contents.
Complex disintegration of the coating of modified or enteric-coated products—for example, the partial opening and splitting apart (like a clamshell) or incomplete opening of the shell accompanied by the release of air bubbles and excipients.
Presence of large floating particles or chunks of the dosage unit.
Observation of the disintegration rate (e.g., percentage reduction in size of the dosage unit within a certain timeframe).

- *IN VITRO-IN VIVO CORRELATION*
  - Dissolution methods can also be utilized to support sophisticated in vitro, in vivo correlation in vitro, in vivo correlation studies in collaboration with Pharmacokinetics. The advantage of the correlations could negate the need for additional clinical studies in the event of a formulation change.
  - Bio-relevant medium, is a medium that has some relevance to the in vivo performance of the dosage unit.
  - Choice of a bio-relevant medium is based on: A mechanistic approach that considers the absorption site, if known

    Convolution and de-convolution studies
    Dissolution correlation with PK profiles
    Whether the rate-limiting step to absorption is the dissolution or permeability of the compound.

- *BCS CLASSIFICATION SYSTEM*
  - The Biopharmaceutics Classification System influences the type of dissolution method to develop and is briefly summarized as follows:

    BCS Class I—high solubility, high permeability drug substance
    BCS Class II—low solubility, high permeability drug substance
    BCS Class III—high solubility, low permeability drug substance
    BCS Class IV—low solubility, low permeability drug substance

Table 3.2 BCS classification versus IV/IVC expectation

Class	Solubility	Permeability	IV/IV correlation expectation
I	High	High	Unlikely
II	Low	High	May be possible
III	High	Low	Unlikely
IV	Low	Low	Most probable

- *Categories of In-Vitro/In-Vivo correlation*
  - Category A—Use of mathematical models (deconvolution) to demonstrate a point to point relationship between in vitro dissolution and the in vivo absorption rate, these curves are usually superimposable.
  - Category B—Comparison of the mean in vitro dissolution time to the mean in vivo dissolution time. This not a point to point correlation, as in category A, Therefore many curves can produce similar mean in vivo dissolution times.
  - Category C—This represents a single point correlation at an in vitro dissolution time point to a pharmacokinetic parameter (AUC, Cmax, Tmax), this category does not reflect the complete shape of the plasma concentration time curve (Table 3.2).

3.2.3 Systematic Approach to Method Validation

Method Validation is a regulatory requirement for any method used for GMP analysis. There is FDA guidance's and ICH guidelines which describe detailed requirement for method validation. However, before one can apply the FDA guidance's and the ICH guidelines there are certain activities that must be identified such as [7]:

- The methods critical performance characteristics for validation
- Main objective of the validation for its intended purpose.
  - Pre-clinical/In-process
  - Clinical
  - Specifications
  - Technology Transfers
- Statistical evaluation of validation data should be performed and documented.

Validation is a continuous process which begins in early development and continues to commercialization and beyond. The level of validation increases throughout the development process, starting with minimal characteristics and then following with a full validation in compliance with the FDA guidance's and ICH guidelines [8].

The validation protocol should identify the criteria for each validation characteristic. Examples of criteria which are generally accepted are as follows:

Assay Criteria

- Accuracy target criteria are ±50 % of the specification limits.
  - e.g., For a specification 90–110 %, the target is 95–105 %
- Precision criteria are such that approximately ±2 SD are within the target Limits.
  - e.g., ±3 % for accuracy target limits of 95–105 %, ±5 % for target limits of 90–110 %
- Linearity/Range criteria are from 50–150 % of the accuracy target concentration, utilizing five concentration levels. Acceptance of linearity is usually based on obtaining:
  - Correlation Coefficient ≥0.99
  - Slope, report results
  - Intercept not significantly different from zero
  - %Bias (residuals) within ±2 SD of the precision value at each concentration level using a linear regression line.
  - If a single point standard is being utilized than %bias from the single point to the theoretical concentrations used in the calibration curve must meet the same criteria.
- Specificity criteria requires no interference from excipients, impurities, and degradation products
- Force degradation studies in acid, base, oxidative, thermal, and light (UV, visible)
  - Mass Balance, the accountability for total mass ≥95 %
- Robustness criteria should evaluate reasonable variation of the methods parameters to encompass random error and the impact on the methods control
  - Robustness results are used to establish or modify system suitability criteria.

Impurities and Degradation products Criteria

- Linearity—Should be demonstrated for the total range of concentrations in the method relative to a single point standard or a calibration curve.
  - Linearity for drug product should be in the placebo matrix.
  - Concentrations from LOQ—150 % of the impurity/degradation products specification

- %Bias within ±2 SD of precision value

- Accuracy—Target is the specification limit ±20 %
  - e.g., For specification 0.5 %, target is 0.5 ± 0.1 %
  - %Bias within ±20 % at each concentration level from the linearity can be considered the accuracy determination.

- Precision—The value should be consistent with the ICH Q2R(1) guideline for repeatability and intermediate precision-RSD ≤10 %.
  - For Unavailable impurities/degradation products use API as surrogate
  - For available impurities/degradation products six replicates at 100 % of the specification level meeting the ICH criteria of RSD ≤10 %
  - Intermediate precision required at phase IIb, phase III

- Quantitation Limit—50 % of the specification for each impurity/degradation product. S/N ≥ 10:1. Variation within ±20 % of the limit
  - For Unavailable impurities/degradation products use API as surrogate
  - Limit of quantitation (LOQ) has to be in the linear range

- Detection Limit—S/N ≥ 3 for each impurity/degradation product
- Specificity-Verify peak purity for the method by, DAD, LC/MS, or other suitable means
  - Forced degradation (thermal, acid, base, oxidative, Light)
  - Spike known impurities and degradation products if available

- Robustness—Comply with ICHQ2R (1)
- Residual Solvent Criteria
  - Validation same as for API impurity
  - Forced degradation in not performed

- Identification
  - HPLC Specificity same as in assay and impurities/degradation products
  - Spectral Specificity—Comparison to Reference Standard and other structurally similar compounds which may be present.
  - If spectra not distinguishable than transformation techniques such as, 2nd derivative spectroscopy may be used
  - Combination of two independent methods such as chromatography and UV spectroscopy may be used

An example of the criteria for validation characteristics at different stages of development are given below (Tables 3.3, 3.4, 3.5, 3.6, 3.7).

Table 3.3 API assay

Characteristics	GLP	Phase I	Phase II	Phase III/Registration
Accuracy	Inferred once linearity, precision, and specificity established	Compare to reference standard with three replicates at 100 % level	Compare to reference standard with three replicates at 100 % level	Compare to reference standard with three replicates at 80 and 150 % levels, with n = 6 at 100 % level, prepared as per method
Precision	Repeatability: one level/ three replicates	Repeatability: 100 % level, three replicates (use accuracy samples)	Repeatability: 100 % level, three replicates (use accuracy samples)	Repeatability and intermediate precision: six replicates at 100 % level
Range	From linearity	From linearity	From linearity	80–150 % supported by linearity, accuracy
Linearity	Five levels, one replicate/ level at 80–120 % can be supported from accuracy	Five levels, one replicate/ level at 80–120% can be supported from accuracy	Five levels, three replicate/ level at 50–120% can be supported from accuracy	Five levels, three replicate/level at 50–120 % can be supported from accuracy
Quantitation limit	–	–	–	–
Detection limit	–	–	–	–
Specificity	Verify peak purity by method development, DAD, LC/MS, or other suitable means	Verify peak purity by method development, DAD, LC/MS, or other suitable means Forced degradation	Verify peak purity by method development, DAD, LC/MS, or other suitable means Forced degradation	Verify peak purity by method development, DAD, LC/MS, or other suitable means Forced degradation Spike known impurities and degradation products, if available
Robustness	Established through method development	Established through method development	Established through method development	Comply with ICH

Table 3.4 API impurities/Degradation products

Characteristics	GLP	Phase I	Phase II	Phase III/Registration
Linearity	Not validated	Five levels, three replicate/level at LOQ-150 % of specification or LOQ-150 % of assay target, using API as a surrogate	Five levels, three replicate/level at LOQ-150 % of specification or LOQ-150 % of assay target, using API as a surrogate	Five levels, three replicate/level at LOQ-150 % of specification for each available impurity and degradation product
Accuracy	Not validated	Compare to reference standard with three replicates at LOQ (0.05 %) using API as a surrogate	Compare to reference standard with three replicates at LOQ (0.05 %) using API as a surrogate	Compare to reference standard at five levels with three replicates at LOQ-150 %, and n = 6 at 100 % of specification for each available impurity and degradation product
Precision	Not validated	Repeatability: establish at LOQ (0.05 %) using API as a surrogate three replicates (use accuracy samples)	Repeatability: establish at LOQ (0.05 %) using API as a surrogate three replicates (use accuracy samples)	Repeatability and intermediate precision: six replicates at 100 % of specification level for each available impurity and degradation product
Range	Not validated	From linearity	From linearity	LOQ-150 % of specification for each available impurity and deg product supported by linearity, accuracy
Quantitation limit	Not validated	Establish at LOQ $S/N \geq 10$ or (0.5 %) using API as a surrogate	Establish at LOQ $S/N \geq 10$ or (0.05 %) using API as a surrogate	$S/N \geq 10$ or 0.05 % or, per ICH, at 50 % of specification for each available impurity and deg product
Detection limit	Not validated	$S/N \geq 3$ or (0.1) using API as surrogate or per ICH	$S/N \geq 3$ or (0.01) using API as surrogate or per ICH	$S/N \geq 3$ or (0.01) for each available impurity/degradation product or per ICH

(continued)

Table 3.4 (continued)

Characteristics	GLP	Phase I	Phase II	Phase III/Registration
Specificity	Not validated	Verify peak purity by method development, DAD, LC/MS, or other suitable means Forced degradation	Verify peak purity by method development, DAD, LC/MS, or other suitable means Forced degradation	Verify peak purity by method development, DAD, LC/MS, or other suitable means Forced degradation Spike known impurities and degradation products
Robustness	Not validated	Established through method development	Established through method development	Comply with ICH

Table 3.5 Drug product assay

Characteristics	GLP	Phase I	Phase II	Phase III/Registration
Linearity	Five levels, one replicate/level at 80–120 %	Five levels, one replicate/ level at 80–150 % (70–130 % for CU as appropriate)	Five levels, three replicate/ level at 80–150 % (70–130 % for CU as appropriate)	Five levels, three replicate/level at 80–150 % (70–130 % for CU as appropriate)
Accuracy	Inferred once linearity, precision, and specificity established (As needed, e.g., for suspensions, perform suitable spiking experiments)	Compare to reference standard with three lowest drug load, spiked placebo three replicates at 100 % level	Compare to reference standard with three lowest drug load, spiked placebo three replicates at 100 % level	Compare to reference standard with three spiked placebo replicates at 80 and 150 % levels, and n = 6 at 100 % level, prepared as per method (70, 100, 130 % for CU as appropriate)
Precision	Repeatability: one level/three replicates	Repeatability: 100 % level, three replicates (use accuracy samples)	Repeatability: 100 % level, three replicates (use accuracy samples)	Repeatability and intermediate precision: six spiked placebo replicates at 100 % level and six authentic final formulation replicates
Range	From linearity	From linearity	From linearity	80–150 % supported by linearity, accuracy (70–130 % for CU as appropriate)
Quantitation limit	–	–	–	–
Detection limit	–	–	–	–
Specificity	Verify peak purity by method development, DAD, LC/MS, or other suitable means	Verify peak purity by method development, DAD, LC/MS, or other suitable means Forced degradation	Verify peak purity by method development, DAD, LC/MS, or other suitable means Forced degradation	Verify peak purity by method development, DAD, LC/MS, or other suitable means Forced degradation Spike known degradation products
Robustness	Established through method development	Established through method development	Established through method development	Comply with ICH

Table 3.6 Drug product degradation products

Characteristics	GLP	Phase I	Phase II	Phase III/Registration
Linearity	Not validated	Five levels, one replicate/level at LOQ-150 % of specification or LOQ-150 % of assay target, using API as a surrogate	Five levels, three replicate/level at LOQ-150 % of specification or LOQ-150 % of assay target, using API as a surrogate	Five levels, three replicate/level at LOQ-150 % of specification for each available degradation product
Accuracy	Not validated	Three lowest drug load spiked placebo replicates at LOQ (0.1 %) using API as a surrogate	Three lowest drug load spiked placebo replicates at LOQ (0.1 %) using API as a surrogate	Compare final dosage forms to reference standard with three replicates at LOQ, 150 %, and n = 6 at 100 % of specification for each available degradation product
Precision	Not validated	Repeatability: three lowest drug load spiked placebo three replicates at LOQ (0.1 %) using API as a surrogate (use accuracy samples)	Repeatability: three lowest drug load spiked placebo three replicates at LOQ (0.1 %) using API as a surrogate (use accuracy samples)	Repeatability and intermediate precision: six replicates at 100 % of specification level for each available degradation product
Range	Not validated	From linearity	From linearity	LOQ-150 % for each available degradation product supported by linearity, accuracy
Quantitation limit	Not validated	Establish at LOQ S/N $\geq$ 10 (0.5 %) using API as a surrogate	Establish at LOQ S/N $\geq$ 10 (0.1 %) using API as a surrogate	0.1 % or, per ICH, at 50 % of specification for each available degradation product

(continued)

Table 3.6 (continued)

Characteristics	GLP	Phase I	Phase II	Phase III/Registration
Detection limit	Not validated	S/N $\geq$ 3 or (0.1 %) using API as surrogate or per ICH	S/N $\geq$ 3 or (0.05 %) using API as surrogate or per ICH	S/N $\geq$ 3 or (0.05 %) for each available degradation product or per ICH
Specificity	Not validated	Verify peak purity by method development, DAD, LC/MS, or other suitable means forced degradation	Verify peak purity by method development, DAD, LC/MS, or other suitable means forced degradation	Verify peak purity by method development, DAD, LC/MS, or other suitable means. forced degradation Spike known degradation products
Robustness	Not validated	Established through method development	Established through method development	Comply with ICH

Table 3.7 Water (KF), GC-residual solvents, identity

Characteristics	GLP	Phase I	Phase II	Phase III/Registration
Water (*KF*)	Use compendial method USP <921>	Use compendial method USP <921>	Use compendial method USP <921> evaluate parameters which differ from compendial method (stir speed, sample size, etc.) to verify method is appropriate for your dosage form	Use compendial method USP <921> Comply with USP <1225>, <1226> validation/ verification of compendial methods for your dosage form
GC-residual solvents	Not validated	Use loss on drying or Use compendial method USP <467>	Use compendial method USP <467> or in house method not validated	Use compendial method USP <467> or in-house method and validate per ICH or USP <1225>, <1226>
Identity	Not validated	Specificity and comparison to reference standard	Specificity and comparison to reference standard	Specificity and comparison to reference standard

3.2.4 Analytical Technology Transfer Process

- General transfer process overview (Fig. 3.7)

  Categories of Method Transfers

- Transfer Waiver
  - Very simple test (e.g., appearance, tablet weight)
  - Method already in Receiving Lab (e.g., water KF, pH, compendial or extensive experience). Requires Phases I and IV
- Method Qualification
  - For qualitative methods (e.g., IR-ID, XRD) and some quantitative methods.
  - Generate acceptable results in Receiving Laboratory. Requires Phases I, II, and IV.
- Co-Validation Transfer
  - Concurrent real time testing in sending and receiving laboratory
  - Real time testing in the receiving laboratory compared to historical results developed in the sending laboratory. Requires Phases I through IV
- Transfer through Validation
  - Requires Phases I through IV

Four Phases of the Method Transfer Process

- Phase I: Document Review
  - Obtain copy of the method validation or qualification report.
  - Obtain copy of the analytical test method (approved or draft).
  - Prepare intent to transfer document.

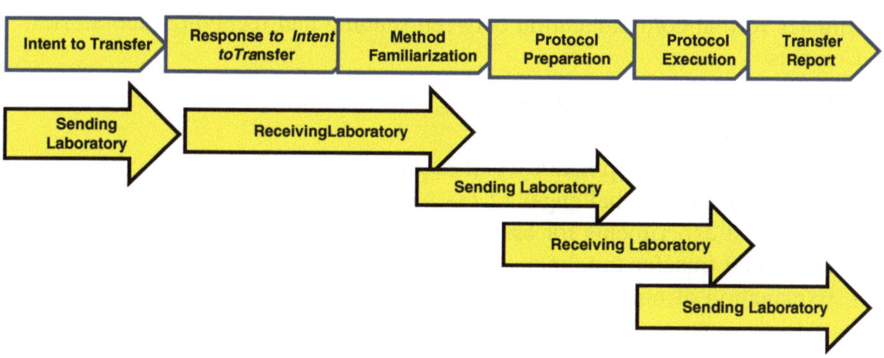

Fig. 3.7 Transfer process

- Phase II: Method Familiarization Study (Training and demonstration of the analytical technology in the receiving laboratory)
  - The receiving laboratory analysts must be trained on the test method procedures by a qualified laboratory analyst from the Sending laboratory, or demonstrate competency.
  - Minor changes to the test method should be documented and as a planned deviation with QA approval.
  - Provide samples and/or standards that are to be tested in both laboratories to establish the adequacy of implementation of the Analytical Methods.
  - Results from the training and testing is examined and a determination if the receiving laboratory is qualified to perform the analytical method.
- Phase III: Validation/Co-Validation
  - Prepare the Method Transfer Protocol incorporating the information from the feasibility study.
  - Provides the required samples to the sending and receiving laboratories.
  - The analysts perform the testing as described in the Method Transfer Protocol.
  - The data is analyzed as required by the protocol and determines if the acceptance criteria have been met.
  - If the criteria have not been met, an investigation is performed by the receiving laboratory. The results from the investigation will dictate the further actions to be taken.
- Phase IV: Preparation of the Final Documents and Approvals
  - All investigations and/or deviations are documented and approved.
  - A Method Transfer Summary Document is written by the receiving laboratory and approved by both the sending and receiving laboratories and Quality Assurance.

Intent to Transfer

- Document prepared by sending laboratory
- Initiates formal communication between sending and receiving laboratories
- Contents
  - Method description/procedure
  - Validation documentation
  - Detailed list of equipment/special requirements
  - Specific technical challenges of the method
  - Suggested experiments for assay familiarization
  - Requested transfer timeline
  - Explanation of non-standard terms or equations

Method Familiarization

- Opportunity to explore method performance/special issues prior to execution of the transfer protocol
- Opportunity to train and qualify analysts in the methods to be performed (when applicable)
- Permits receiving laboratory to assess if method is consistent with local practices and suggest changes
- Ensures that method transfer experiments assess method performance rather than analyst experience/training
- Experiments targeted to confirm critical method performance characteristics.

Method Transfer Protocol

- Prepared by the Sending Laboratory in collaboration with the Receiving Laboratory
- Contains: transfer study design, specific lot, acceptance criteria, number of results needed, sample requirements
- Results from Sending Lab may be available (historical)—must assure are appropriate
- Approvals: Sending Lab, Receiving Lab, and QA.

Method Transfer Report

- Prepared by the receiving laboratory with input from the sending laboratory
- Contains: results, statistical analysis, assessment versus acceptance criteria
- Approvals:
  - Sending Laboratory management
  - Receiving Laboratory management
  - Receiving Laboratory QA.

Acceptance Criteria

- Precision
  - Upper 95 % Confidence Interval of the %RSD cannot be exceeded at either site
- Comparison of Means
  - The mean values obtained from the transfer study must be contained within the 95 % confidence interval of means. Analytical method transfer process (Fig. 3.8).

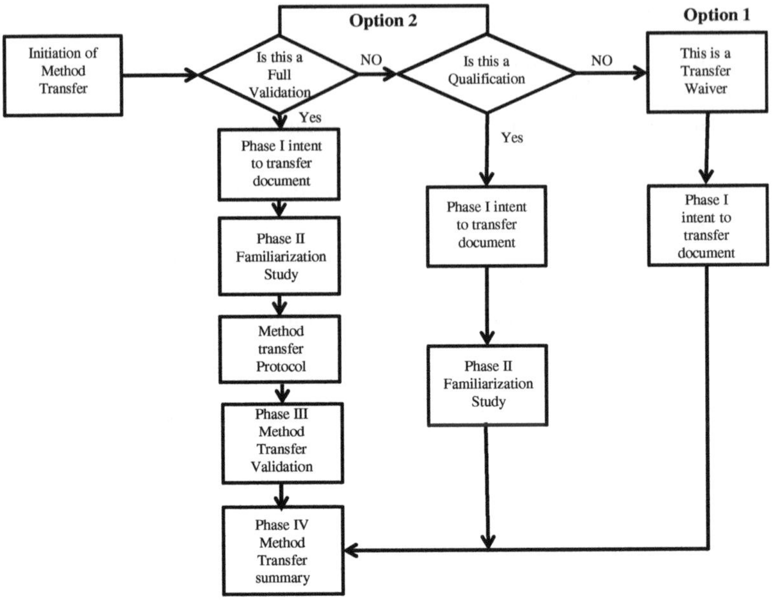

Fig. 3.8 Analytical method transfer process

3.3 Departmental Processes

3.3.1 Specification Development Process

To control the quality, purity, identity, strength, and stability of Drug Substance, Drug Product, Raw materials, Intermediates, Process impurities, Degradation products and non-active ingredients specifications must be determined and complied with over the life of the Material. An effective specification development process should be implemented utilizing the following considerations.

- Utilize a harmonized approach for proposing specifications in drug substances and drug products under various phases of development
- Acceptance Criteria should be consistent with regulatory guidance's and ICH guidelines and Manufacturing Experiences [7]
- During Pre-marketing Phases it is the aim to follow the spirit of the guidance's while at the same time considering higher thresholds and broader specifications for material used in early development. The ICH concept of Qualification will be followed to ensure that there is no safety risks at the limit of the specification proposed.

Description and Identity

- Description
  - Physical State should be described as a solid, liquid, or gas
  - The dosage form should be described as simple as possible for visual recognition (tablet, capsule, scored, opaque, coated, color (qualitatively), etc.)
  - Organoleptic properties should not be used in specification
- Identity
  - Single versus Multiple ID specification

    Multiple ID is usually required for registration or if a single ID is not selective.
    Single selective ID is acceptable during the development stages.
    Method should be selective for related substances.
    Method selectivity should be based on structural information.
  - Solid State

    Identity specifications are required for chirality and Polymorphism

Assay

- Assay Specificity
  - Assay method should be stability indicating
  - Assay method should produce data that is consistent with data from impurities/degradation products (Mass Balance)
- Chirality
  - Chiral assay is required for compounds with a single chiral center
  - No Chiral assay is required for drug product, if demonstrated that racemization does not occur during manufacturing
  - Optical rotation can be used in place of a chiral assay, if the specific rotation is sufficiently high to provide adequate sensitivity
  - Compounds with multiple chiral centers may utilize non chromatographic tests if properly validated
- Pharmaceutical Salts
  - Counter-ion specifications are not required unless it is demonstrated that the counter-ion is critical to the performance of the drug
- Specification setting
  - All data from final process should be utilized, with consideration of data from pilot and development lots
  - Assay specification should be data driven but should not supplant safety considerations, tighter in house limits should utilized

Impurities

- Nature of Impurities
  - Impurities are classifies as Organic, Inorganic, Residual Solvents
  - Foreign Matter of Allergens should not be considered as an impurity, these are considered GMP issues and not specification issues
- Specification Parameters
  - Drug substance specifications should include relevant impurities, such as starting materials, pivotal and penultimate intermediates, residual solvents, unknown impurities.
  - Enantiomers and polymorphs should not be included as drug substance specification for impurities, they are treated as separate specifications.
  - Polymorph specification should only be included if it is known to affect the drug product attributes.
  - Drug product specifications should include enantiomers, stereoisomers, and drug substance impurities that are found to be degradation products. Solvents used in the drug product manufacturing may also require specifications.
  - Specifications on degradation products should only be required if observed under ICH conditions, however if no degradation product are observed a default specification for total unknown impurities will be included.
- Setting Specifications
  - Specifications are set to assure product safety and monitor quality.
  - Manufacturing capabilities should be considered when setting specifications, but should not supplant safety.
  - Statistical considerations which take into consideration the level of the impurity qualified, manufacturing capabilities, and analytical variability should be utilized for setting specifications.
  - Specifications which are above the manufacturing capability and below the qualification limit can be allowed.
  - USP rounding rules should be applied.
  - Specifications for total impurities should not be determined utilizing measurements less than the analytical LOQ specification.

Dissolution

- When is dissolution testing required?
  - To determine chemical and formulation differences (Quality Control)

    Solubility
    Intrinsic Dissolution
    Particle Size
    Excipient Differences
    Manufacturing Differences

- To address product performance (in vivo/in vitro correlation)

  Permeability
  Bioavailability
  Partition Coefficients

- Dissolution versus Disintegration
  - Highly soluble compound
  - Rapid dissolution—≥85 % in 15 min
  - Must establish relationship between dissolution and disintegration by acquiring sufficient data during development
- Categories of In-Vitro/In-Vivo correlation
  - Category A-Use of mathematical models (deconvolution) to demonstrate a point to point relationship between in vitro dissolution and the in vivo absorption rate, these curves are usually superimposable.
  - Category B- Comparison of the mean in vitro dissolution time to the mean in vivo dissolution time. This is not a point to point correlation, as in category A, Therefore many curves can produce similar mean in vivo dissolution times.
  - Category C- This represents a single point correlation, which is one in vitro dissolution time point to one pharmacokinetic parameter (AUC, Cmax, Tmax), this level does not reflect the complete shape of the plasma concentration time curve (Table 3.8).

In-Vitro/In-Vivo Correlations for Immediate Release Products Based on Biopharmaceutical Classification (Table 3.9).

Table 3.8 Chemical classification table

I Solubility/Permeability Hi/Hi 1. Initial dissolution testing 2. Disintegration specification	II Solubility/Permeability Lo/Hi 1. Dissolution testing specification
III Solubility/Permeability Hi/Lo 1. Initial dissolution testing 2. Disintegration specification	IV Solubility/Permeability Lo/Lo 1. Dissolution testing specification

Table 3.9 IV/IV correlations expectations

Class	Solubility	Permeability	IV/IV correlation expectation
I	High	High	Unlikely
II	Low	High	May be possible
II	High	Low	Unlikely
III	Low	Low	Most probable

- Profile versus Single Point
  - Immediate release formulations should set a single point specification, the point should not be less than 30 min.
  - Extended release formulations should set 3–5 point specification

    First point (20–30 % released), second point (50–80 % released), final point (at complete release)

  - Extended release formulation specification should include multiple pH's if appropriate
- Specification should be data driven and should take into consideration the cumulative knowledge gained during development
  - Statistical Techniques should be utilized where there is sufficient data to apply them
  - USP type S2 testing is recommended at release and stability specification
- Dissolution Method
  - Apparatus should be USP type
  - Paddle speed, 50–100 rpm
  - Dissolution Media

Aqueous system (pH 1.2–6.8) > Surfactants/enzyme system > Co-Solvent system Extended release should be tested across pH ranges of 1, 4, 5, 7 if no pH dependence then develop in aqueous system

- Method sensitivity should be at least 3 × lower then sink conditions

Proposing Specifications at various stages of development

- For pre-clinical Toxicology Studies, a balance is required between the desire to qualify impurities and the need to preserve the integrity of the toxicology studies for determining the characteristics of the drug substance itself. An Example is show in Table 3.10:

Table 3.10 Specification for various stages of development

Toxicology studies	Individual impurities	Total impurities
Pre-clinical genetic toxicology studies	NMT 0.5 % any single impurity	NMT 2 % total
All other toxicology studies during exploratory Development	NMT 1 % any single impurity	NMT 5 % total

Table 3.11 Drugs for human use

DOSE (per day)	Individual impurities (Phase 1) (%)	Total impurities (Phase 1)
<10 mg	NMT 1	Report results
10–100 mg	NMT 0.5	Report results
101 mg–2 g	NMT 0.5	Report results
>2 g	NMT 0.2	Report results

- Specifications recommended for API lots used in Clinical trials is shown in Table 3.11
- New Impurities not tested may be consider at these levels with some information (structure, risk data base, etc.). If no information available NMT 0.2 % is acceptable for all doses/per day
- In support of Early Development, specifications for simple lab scale formulations are recommended.

Table 3.12 API in a bottle

Test	Method	Limit
Potency	Fill weight[a]	The active ingredient may range from 90.0 to 110.0 % of the labeled amount, calculated as the average fill weight multiplied by the purity of the bulk drug lot
Appearance	Visual	Meets description
Identification	HPLC[b]	The retention time of the major peak in the chromatogram of the sample preparation corresponds to that of the active ingredient peak in the chromatogram of the reference standard preparation
Dose uniformity	Weight variation[a]	Meets current USP requirements
Clarity of solution	Visual	Clear and free of particulate matter

[a] Using fill weight data from manufacturing record
[b] Spectroscopic method may be utilized

Table 3.13 API in a capsule

Test	Method	Limit
Potency	Fill weight[a]	The active ingredient may range from 90.0 to 110.0 % of the labeled amount, calculated as the average fill weight multiplied by the unity of the bulk drug lot
Appearance	Visual	Meets test
Identification	HPLC[b]	The retention time of the major peak in the chromatogram of the sample preparation corresponds to that of the active ingredient peak in the chromatogram of the reference standard preparation
Dose uniformity	Weight variation[a]	Meets current USP requirements
Dissolution	Compendial apparatus	Report results Internal limit appropriate to dosage form, or data review, may be utilized. Evaluate global compendial requirements for acceptance criteria
Disintegration	Compendial apparatus	Consider for in-process testing

[a] Using fill weight data from manufacturing record; assumes hand fill manufacture
[b] Spectroscopic method may be utilized

- *API in Bottle*
- *API in Capsule*
- Sterile Solution

- Generic specifications are proposed as follows (Table 3.12):
- In the proposed generic specifications, Constitution can be handled as instructions to the clinic without a specific test in the specifications, but information will be needed to support quantitative recovery of drug from the container.
- Constituted stability or microbial quality will not be issues if the clinic is instructed to dose immediately after constitution.
- The API in a bottle should be put on a abbreviated stability study, since the API stability program can support the stability of the API in a bottle.

Table 3.14 Sterile solution (for oral or IV use)

Test	Method	Limit
Potency	HPLC	The active ingredient may range from 90.0 to 110.0 % of the labeled amount
Appearance	Visual	Meets test
Identification	HPLC	The retention time of the major peak in the chromatogram of the sample preparation corresponds to that of the active ingredient peak in the chromatogram of the reference standard preparation
Dose uniformity	Compendial	Injection: meets USP requirements for volume of injection Oral: multi dose: meets USP requirements for minimum fill, deliverable volume Unit dose: meets USP requirements for dose uniformity
Microbial quality		
Sterility	Compendial	Meets test
Endotoxins[a]	Compendial	Limit based on dose
Particulate matter[b]	Compendial	Meets current USP requirements for subvisible particulates
Impurities (Total and individual degradation products)	HPLC	Report results The potential need for a numeric limit is based on evaluation of batch data, stability data, and safety
pH	Potentiometric	Report results A numeric limit should be considered if pH is stability indicating. In addition to potency/impurities, consider e.g., solubility, in-process control limits. Also consider the potential for physiological implications, e.g., pain on injection

[a] Not required for oral administration
[b] Not required for oral, SQ, or IM administration

Table 3.15 Drug substance—investigational phase 1 and 2

Test procedure	Registration acceptance criteria	Internal acceptance criteria	Comments
Preliminary specifications (Phase 1)			
Assay	Limit supported by batch data and stability data	Limit supported by batch data and stability data	Results may be expressed on the anhydrous, solvent excluded basis; "as is"; or both
Description or appearance	Descriptive text	Descriptive text	Organoleptic tests should be avoided
Identification			
A	Positive Infrared absorption spectrum conforms to that of standard	Positive Infrared absorption spectrum conforms to that of standard	Identification A and B may be IR and HPLC, respectively. Other structure specific methods may be utilized instead of IR
B	–	Positive Retention time of major sample peak corresponds to that of standard	
Impurities			
Total	Limit based on dose, qualification level, and phase of clinical development	Limit based on dose, qualification level, and phase of clinical development	A limit of reports results for total impurities is acceptable for phase 1 registration acceptance criteria
Individual	Limit based on dose, qualification level, and phase of clinical development (see suggested criteria)	Limit based on dose, qualification level, and phase of clinical development (see suggested criteria)	
Impurities Inorganic	–	Limited support by batch data and pharmacopeial precedents	The need for inorganic impurities tests, such as sulfated ash and heavy metals, should be evaluated. Check synthesis for use of a catalyst
Residual solvents[8a,10] Total Individual	– Report results	Report results Report results	When numeric limits are required, ICH limits can be the basis for discussion
Specific rotation	Report results	Report results	Need for a chiral drug

(continued)

Table 3.15 (continued)

Test procedure	Registration acceptance criteria	Internal acceptance criteria	Comments
Particle size	–	Report results	This test should be considered for a solid or suspension dosage form
Water	Report results	Report results	A Karl Fischer method is often described; specify the method, e.g., volumetric or coulometric. Loss on dying could be considered early in development. A numeric limit should be considered if stability of the compound or subsequent processing is sensitive to the presence of water
Microbial quality	–	Limit appropriate to dosage form	Microcount/pathogens, sterility (e.g., for a bulk drug manufactured as sterile), or endotoxins (e.g., for a bulk drug used in an injectable dose form). Evaluate global compendial requirements for methods and limits
Impurities—enantiomeric	Report results	Report results	Needed for a chiral drug. The potential need for a numeric limit is based on evaluation of batch data, stability data, and safety
Solid state forms	Properties needed to demonstrate process control should be considered	Properties needed to demonstrate process control should be considered	Needed when a polymorph or solvate is present. XRD is typically utilized

Table 3.16 Drug substance—phase 3 and market product

Test procedure	Registration acceptance criteria	Internal acceptance criteria	Comments
Assay	Limit supported by batch data and stability data	Limit supported by batch data and stability data	Results are typically expressed on the anhydrous, solvent excluded basis; "as is"; or both. For a chiral drug, an enantioselective determination may be needed. A lower limit of not less than 98.0 % is not uncommon for the market application, A microbiological assay may be needed for antibiotics
Description or appearance	Descriptive text	Descriptive text	Organoleptic tests should be avoided. Degree of coloration may also be considered, especially for drug to be used in injectable dosage forms
Identification			Identification A and B may be IR and HPLC, respectively
A	Infrared absorption spectrum conforms to that of standard	Infrared absorption spectrum conforms to that of standard	Other structure specific methods may be utilized instead of IR. HPLC may be used for lot release. For a salt, need to also include identification testing for the counter ion. For a chiral drug, a chiral specific method may be needed (consider chiral HPLC or specific rotation)
B	Positive Retention time of major sample peak corresponds to that of standard	Positive Retention time of major sample peak corresponds to that of standard	
Impurities			Safety information is needed for the qualification of specified impurities above ICH threshold limits
Total	Limit supported by batch data and stability data	Limit supported by batch data and stability data	
Specified Unspecified	Limit based on the qualification level and supported by batch data and stability data Human health: NMT 0.1 %	Limit based on the qualification level and supported by batch data and stability data Human health: NMT 0.1 %	
Impurities* Enantiomeric	Limit supported by batch data and stability data	Limit supported by batch data and stability data	May be needed for a chiral drug. Data collected throughout development may reduce the need of a chiral purity specification to chiral identity specification. Safety information is needed

(continued)

Table 3.16 (continued)

Test procedure	Registration acceptance criteria	Internal acceptance criteria	Comments
Impurities[13,21] Inorganic	Limit supported by batch data and pharmacopeia precedents	Limit supported by batch data and pharmacopeia precedents	The need for inorganic impurities tests, such as sulfated ash and heavy metals, should be evaluated. Sulfated ash may not be applicable to a salt. Check synthesis for use of a catalyst
Residual solvents Total	–	Limit supported by batch data and ICH classification	Testing should be performed for residual solvents when production or purification processes are known to result in the presence of such solvents. It is only necessary to test for solvents that are used or produced during manufacture or purification. A requirement a registered limit for total residual solvents should be considered
Individual	Limit supported by batch data and ICH classification	Limit supported by batch data and ICH classification	
Specific rotation	Limit supported by batch data and stability data, Must be supported with chiral data	Limit supported by batch data and stability data. Must be support with chiral data	Include data in file; limit may not be needed if stereo chemical purity is determined
Physicochemical properties	Limit supported by batch data	Limit supported by batch data	The need for these tests is determined by the physical nature of the drug and its intended use, e.g., pH of aqueous solution, melting point, refractive index, and density/bulk density
Particle size	Limit supported by batch data bioavailability, or stability data	Limit supported by batch data bioavailability, or stability data	This test should be considered for a solid or suspension dosage form. Manufacturability of the dosage form may also be a factor in the control of particle size
Solid state forms	Properties needed to demonstrate process control should be considered	Properties needed to demonstrate process control should be considered	Needed when a polymorph or solvate could affect drug product performance, safety, efficacy, bioavailability, or stability. A quantitative limit may be needed
Water	Limit supported by batch and stability data	Limit supported by batch and stability data	A Karl Fischer method is usually described; specify the method, e.g., volumetric or coulometric
Microbial quality	Limit based on compendia requirements, appropriate to dosage form	Limit based on compendia requirements, appropriate to dosage form	Micro count/pathogens, sterility (e.g., for a bulk drug manufactured as sterile), or endotoxins (e.g., for a bulk drug used in an injectable dosage form). Evaluate global compendia requirements for methods and limits

Table 3.17 Solid oral drug products, investigational phase 1 and 2 (Includes tablets, coated tablets, and hard filled capsules)

Test procedure	Registration acceptance criteria	Internal acceptance criteria	Comments
Assay	95.0–105.0 % of label	95.0–105.0 % of label	Shelf life limit may be broader, with supporting data. Justification of a broader time of release limit may also be possible. A limit of 90.0–110.0 % of label is usually acceptable for the U.S. and many other markets. A limit of 95.0–105.5 % of label is expected for the EU
Description or appearance	Descriptive text	Descriptive text	Some countries require a very detailed description. Organoleptic tests should be avoided
Identification	Retention time of major sample peak corresponds to that of standard	Retention time of major sample peak corresponds to that of standard	HPLC may be utilized. A second identification test is usually not required at this stage
Dissolution	Report results	Report results	Internal limit appropriate to dosage form, or data review, may be utilized. Evaluate compendia requirements for acceptance criteria
Disintegration	–	Limit appropriate to dosage form	Consider for in-process testing
Hardness/friability/weight/thickness	–	Report results or limit supported by batch data	Consider for in-process testing
Dose uniformity	Meets compendia requirements	Meets compendia requirements	Evaluate compendia requirements for methods and limits
Water	–	Report results	A Karl Fischer method is usually described; specify the method, e.g., volumetric or coulometric. A numeric limit should be considered if product performance is sensitive to the presence of water
Microbial quality	–	Limit appropriate to dosage form	Evaluate global compendia requirements for methods and limits

(continued)

Table 3.17 (continued)

Test procedure	Registration acceptance criteria	Internal acceptance criteria	Comments
Preservative content	Report results	Report results	Consider both antimicrobial preservatives and antioxidants. For antimicrobial preservatives, preservative effectiveness should be demonstrated during development, scale up, and throughout shelf life[20]. The potential need for a numeric limit is based on evaluation of batch data, stability data, and antimicrobial effectiveness data. Evaluate compendia requirements for AET methods and limits
Total degradation products Individual degradation products	Report results Report results	Report results Report results	During development, while degradation mechanisms are being characterized, it is important to evaluate how process impurities and degradation products will be monitored. The potential need for a numeric limit is based on evaluation of batch data, stability data, and safety… Impurities which are controlled by drug substance specification need not be included in drug product specifications
Impurities Enantiomeric	–	Report results	The potential need for a numeric limit is based on evaluation of batch data, stability data, and safety. Impurities which are controlled by drug substance specification need not be included in drug product specifications
Solid state forms	Properties needed to demonstrate process control should be considered	Properties needed to demonstrate process control should be considered	Needed when a polymorph or solvate is present. A surrogate test may be utilized

Table 3.18 Solid oral drug products—phase 3 and market product includes (Tablets, coated tablets, and hard filled capsules)

Test procedure	Registration acceptance criteria[12]	Internal acceptance criteria	Comments
Assay	95.0–105.0 % of label	95.0–105.0 % of label	Shelf life limit may be broader, with supporting data. Justification of a broader time of release limit may also be possible. A limit of 90.0–110.0 % of label is usually acceptable for the U.S. A limit of 95.0–105.5 % of label is expected for the EU. For a chiral drug, an enantioselective determination may be needed. A microbiological assay may be needed for antibiotics
Description or appearance	Descriptive text	Descriptive text	If color changes during storage (i.e., an unavoidable and acceptable change in appearance), a quantitative procedure may be needed. Organoleptic tests should be avoided
Identification			Identification A and B may be IR and HPLC, respectively. Other structure specific methods may be utilized instead of IR. HPLC or UV may be used for lot release
A	Infrared adsorption spectrum conforms to that of standard	Infrared adsorption spectrum conforms to that of standard	
B	Retention time of major sample peak corresponds to that of standard	Retention time of major sample peak corresponds to that of standard	
Total degradation products	Limit supported by batch data and stability data	Limit supported by batch data and stability data	Safety information is needed for the qualification of specified impurities above threshold limits. Qualification thresholds are different for human and veterinary products. For a chiral drug, control of the enantiomer is necessary if it is a degradation product
Unspecified degradation products	Limit based on the qualification level and supported by batch data and stability data	Limit based on the qualification level and supported by batch data and stability data	
Specified degradation products	Human health: Limited based on maximum daily dose and ICH threshold limits, or qualified through safety studies	Human health: Limited based on maximum daily dose and ICH threshold limits, or qualified through safety studies	
Dissolution/ disintegration	Limit supported by batch data, stability data, and bioavailability data	Limit supported by batch data, stability data, and bioavailability data	Evaluate compendia requirements for acceptable criteria

(continued)

Table 3.18 (continued)

Test procedure	Registration acceptance criteria[12]	Internal acceptance criteria	Comments
Hardness/friability/ weight/thickness	Limited supported by batch data	Limited supported by batch data	Consider for in-process testing
Dose uniformity	Meets USP, Ph.Eur., and JP requirements	Meets USP, Ph.Eur., and JP requirements	Evaluate global compendia requirements for methods and limits. Tablets with a functional score must also demonstrate dose uniformity for tablet halves
Water	Limit supported by batch data and stability data	Limit supported by batch data and stability data	A Karl Fischer method is usually described; specify the method, e.g., volumetric or coulometric
Microbial quality	Limit based on compendia requirements	Limit based on compendia requirements	Microcount/pathogens. Evaluate compendia requirements for methods and limits
Preservative content	Limit support by batch data, stability and AET data (for antimicrobial preservatives)	Limit support by batch data, stability and AET data (for antimicrobial preservatives)	Consider both antimicrobial preservatives antioxidants. A limit of 90.0–110.0 % of label at time of release may be expected; shelf life limit may be broader, with supporting data. For antimicrobial preservatives, preservative effectiveness should be demonstrated during development, scale up, and throughout shelf life[20]. Evaluate compendia requirements for AET methods and limits. Some markets a preservative identification test may be required, and HPLC may be adequate. Evaluate compendia requirements for antioxidant method and limits
Solid state forms	Properties needed to demonstrate process control should be considered	Properties needed to demonstrate process control should be considered	Needed when a polymorph or solvate could affect drug product performance, safety, efficacy, bioavailability, or stability–if the form is known to change during manufacture or on stability. A surrogate test is preferred

(continued)

Table 3.18 (continued)

Test procedure	Registration acceptance criteria[12]	Internal acceptance criteria	Comments
Excipient identification	Positive	Positive	A test for identification of colorants is required. Identification of all excipients has be a requirement for drug products sourced from outside of the EU. Evaluate whether this is needed for each dossier, as the requirement appears to be diminishing

Table 3.19 Parenteral drug products—investigational phase 1 and 2 (Includes, sterile solutions, sterile powders, sterile suspensions)

Test procedure	Registration acceptance criteria	Internal acceptance criteria	Comments
Assay	95.0–105.0 % of label	95.0–105.0 % of label	Shelf life limit may be broader, with supporting data. Justification of a broader time of release limit may also be possible. A limit of 90.0–110.0 % of label is usually acceptable for the U.S. A limit of 95.0–105.5 % of label is expected for the EU
Description or appearance	Descriptive text	Descriptive text	Some countries require a very detailed description. For a solution, should be essentially free of visible particulates, and Ph.Eur. Degree of coloration as well as clarity and opalescence tests and limits may also be considered. Organoleptic tests should be avoided
Identification	Retention time of major sample peak corresponds to that of standard	Retention time of major sample peak corresponds to that of standard	HPLC may be utilized. A second identification test is usually not required at this stage
Dose uniformity	Meets compendia requirements	Meets compendia requirements	Evaluate compendia requirements for methods and limits. Compendia fill requirements must also be considered (USP volume of injection,)
pH	Report results	Report results	A numeric limit should be considered if pH is stability indicating. In addition to potency/impurities, consider e.g., solubility, in-process control limits. Also consider the potential for physiological implications, e.g., pain on injection
Microbial quality			
Sterility	Meets test	Meets test	Evaluate compendia requirements for methods and limits
Endotoxins	Limit calculated based on dose and administration route	Limit calculated based on dose and administration route	
Particulate matter	Limit based on compendia requirements	Limit based on compendia requirements	Evaluate compendia requirements for tests and limits. In addition to sub visible particulates, need to include limits for visible particulates and/or clarity of solution

(continued)

Table 3.19 (continued)

Test procedure	Registration acceptance criteria	Internal acceptance criteria	Comments
Water	–	Report results	A Karl Fischer method is usually described; specify the method, e.g., volumetric or coulometric. A numeric limit should be considered if product performance is sensitive to the presence of water
Preservative Content	Report Results	Report Results	Consider both antimicrobial preservatives and antioxidants. For antimicrobial preservatives, preservative effectiveness should be demonstrated during development, scale up, and throughout shelf life[20]. The potential need for a numeric limit is based on evaluation of batch data, stability data, and antimicrobial effectiveness data. Evaluate compendia requirements for AET methods and limits
Particle size distribution	Report results	Limit supported by batch data and stability data	For injectable suspensions, may also need dissolution
Specific gravity	–	Report results	
Rheological properties	–	Limit supported by batch data and stability data	May be needed for relatively viscous drug product
Supplemental specifications (add when available)			
Total degradation products	Report results	Report results	During development, while degradation mechanisms are being characterized, it is important to evaluate how process impurities and degradation products will be monitored. The potential need for a numeric limit is based on evaluation of batch data, stability data, and safety
Individual degradation products	Report results	Report results	
Impurities Enantiomeric	–	Report results	The potential need for a numeric limit is based on evaluation of batch data, stability data, and safety. Impurities which are controlled by drug substance specification need not be included in drug product specifications
Solid state forms	Properties needed to demonstrate process control should be considered	Properties needed to demonstrate process control should be considered	Needed when a polymorph or solvate is present. A surrogate test may be utilized

Table 3.20 Parenteral drug products—phase 3 and marketed product (Includes, sterile solutions, sterile powders, sterile suspensions)

Test procedure	Registration acceptance criteria	Internal acceptance criteria	Comments
Assay	95.0–105.0 % of label	95.0–105.0 % of label	Shelf life limit may be broader, with supporting data. Justification of a broader time of release limit may also be possible. A limit of 90.0–110.0 % of label is usually acceptable for the U.S. A limit of 95.0–105.5 % of label is expected for the EU. For a chiral drug, an enantioselective determination may be needed. A microbiological assay may be needed for antibiotics
Description or appearance	Descriptive text	Descriptive text	Some countries require a very detailed description. If color changes during storage (i.e., an unavoidable and acceptable change in appearance), a quantitative procedure may be needed. For a solution, Ph.Eur. Degree of coloration as well as clarity and opalescence tests and limits may also be considered. Organoleptic tests should be avoided
Identification			Identification A and B may be IR and HPLC, respectively. Other structure specific methods may be utilized instead of IR. HPLC may be used for lot release. For a chiral drug, a chiral specific method may be needed
A	Infrared adsorption spectrum conforms to that of standard	Infrared adsorption spectrum conforms to that of standard	
B	Retention time of major sample peak corresponds to that of standard	Retention time of major sample peak corresponds to that of standard	
Degradation products			Safety information is needed for the qualification of specified impurities above threshold limits. Qualification thresholds are different for human and veterinary products. For a chiral drug, control of the other enantiomer is necessary if it is a degradation product
Total	Limit supported by batch data and stability data	Limit supported by batch data and stability data	
Specified degradation products	Limit based on the qualification level and supported by batch data and stability data	Limit based on the qualification level and supported by batch data and stability data	
Unspecified degradation products	Limited based on maximum daily dose and ICH threshold limits	Limited based on maximum daily dose and ICH threshold limits	Impurities which are controlled by drug substance specification need not be included in drug product specifications

(continued)

Table 3.20 (continued)

Test procedure	Registration acceptance criteria	Internal acceptance criteria	Comments
Dose uniformity	Meets USP, requirements	Meets USP, requirements	Evaluate compendia requirements for methods and limits. Compendia fill requirements must also be considered (USP volume of injection,)
pH	Limit supported by batch data and stability data	Limit supported by batch data and stability data	Consider, e.g., stability potency/impurities, solubility, and in-process control limits. Also consider the potential for physiological implications, e.g., pain on injection
Microbial quality			Evaluate compendia requirements for methods and limits
Sterility	Meets test	Meets test	
Endotoxins	Limit calculated based on dose and administration route	Limit calculated based on dose and administration route	
Particulate matter	Limit based on compendia requirements	Limit based on compendia requirements	Evaluate compendia requirements for sub visible particulates tests and limits. Also need to include limits for visible particulates and/or clarity of solution
Water	Limit supported by batch and stability data	Limit supported by batch and stability data	A Karl Fischer method is usually described; specify the method, e.g., volumetric or coulometric
Preservative content	Limit support by batch data, stability data, and AET data (for antimicrobial preservatives)	Limit support by batch data, stability data, and AET data (for antimicrobial preservatives)	Consider both antimicrobial preservatives and antioxidants. A limit of 90.0–110.0 % of label at time of release may be expected; shelf life limit may be broader, with supporting data. For antimicrobial preservatives, preservative effectiveness should be demonstrated during development, scale up, and throughout shelf life[20]. Evaluate compendia requirements for AET methods and limits. Some markets (e.g., EU) also require a preservative identification test, and HPLC may be adequate. Evaluate compendia requirements for antioxidant method and limits
Extractable	Limit supported by batch data and stability data	Limit supported by batch data and stability data	Safety information will be needed. Consider both specific and non-specific (compendia) tests. Should also consider current draft guidance's on extractables

(continued)

Table 3.20 (continued)

Test procedure	Registration acceptance criteria	Internal acceptance criteria	Comments
Functional test for delivery systems	Limit supported by batch data	Limit supported by batch data	For formulations packaged in pre-filled syringes, auto injector cartridges, or the equivalent
Osmolarity	Limit supported by batch data	Limit supported by batch data	Needs to be considered when tonicity is declared in the labeling
Particle size distribution	Limit supported by batch data and stability data, and dissolution data	Limit supported by batch data and stability data, and dissolution data	For injectable suspensions, may also need dissolution
Specific gravity	Limit supported by batch data and stability data	Limit supported by batch data and stability data	
Rheological properties	Limit supported by batch data and stability data	Limit supported by batch data and stability data	May be needed for relatively viscous drug products. A measure of syringeability should also be considered for suspensions. Related tests, such as droplet size in emulsions, should be considered
Redispensability	Limit supported by batch data	Limit supported by batch data	May be needed for a suspension which settles on storage. Particle size is a factor, and assay/homogeneity may be affected
Reconstruction time	Limit supported by batch data	Limit supported by batch data	
Solid state forms	Properties needed to demonstrate process control should be considered	Properties needed to demonstrate process control should be considered	Needed when a polymorph or solvate could affect drug product performance, safety, efficacy, bioavailability, or stability–if the form is known to change during manufacture or on stability. A surrogate test is preferred
Excipient identification	Positive		Identification of all excipients has been a requirement for drug products sourced from outside of the EU (test provided to be applied if needed, not included in finished product specifications). Evaluate whether this is needed for each dossier, as the requirement appears to be diminishing

Table 3.21 Oral liquid drug products—investigational phase 1 and 2 (Includes oral solutions, suspensions and reconstitutable powders)

Test procedure	Registration acceptance criteria	Internal acceptance criteria	Comments
Assay	95.0–105.0 % of label	95.0–105.0 % of label	Shelf life limit may be broader, with supporting data. Justification of a broader time of release limit may also be possible. A limit of 90.0–110.0 % of label is usually acceptable for the U.S. and many other markets. A limit of 95.0–105.5 % of label is expected for the EU
Description or appearance	Descriptive text	Descriptive text	Some countries require a very detailed description. For a solution, Ph.Eur. Degree of coloration as well as clarity and opalescence tests and limits may also be considered. Organoleptic tests should be avoided
Identification	Positive Retention time of major sample peak corresponds to that of standard	Positive Retention time of major sample peak corresponds to that of standard	HPLC may be utilized. A second identification test is usually not required at this stage
Dose uniformity	Meets compendia requirements	Meets compendia requirements	Evaluate global compendia requirements for methods and limits. Compendia fill requirements must also be considered. (Minimum fill and deliverable volume)
pH	Report results	Report results	A numeric limit should be considered if pH is stability indicating. In addition to potency/impurities, consider e.g., solubility, in-process control limits
Microbial quality	–	Limit appropriate to dosage form	Evaluate global compendia requirements for methods and limits
Preservative content	Report results	Report results	Consider both antimicrobial preservatives and antioxidants. For antimicrobial preservatives, preservative effectiveness should be demonstrated during development, scale up, and throughout shelf life[20]. The potential need for a numeric limit is based on evaluation of batch data, stability data, and antimicrobial effectiveness data. Evaluate global compendia requirements for AET methods and limits

(continued)

Table 3.21 (continued)

Test procedure	Registration acceptance criteria	Internal acceptance criteria	Comments
Dissolution	Report results	Report results	Internal limit appropriate to dosage form, or data review, may be utilized. Evaluate global compendia requirements for acceptance criteria
Rheological properties	–	Limit supported by batch data and stability data	May be needed for relatively viscous drug product
Specific gravity	–	Report results	
Water	–	Report results	A Karl Fischer method is usually described; specify the method, e.g., volumetric or coulometric. A numeric limit should be considered if product performance is sensitive to the presence of water. In addition to potency/impurities, consider, e.g., reconstitution
Supplemental specifications (Add when available)			
Impurities	Report results	Report results	During development, while degradation mechanisms are being characterized, it is important to evaluate how process impurities and degradation products will be monitored. The potential need for a numeric limit is based on evaluation of batch data, stability data, and safety
Total degradation products	Report results	Report results	
Individual degradation products			
Impurities Enantiomeric	–	Report results	The potential need for a numeric limit is based on evaluation of batch data, stability data, and safety
Solid state forms	Positive	Positive	Needed when a polymorph or solvate is present. A surrogate test may be utilized

Table 3.22 Oral liquid drug products—phase 3 and marketed product (Includes oral solutions, suspensions and reconstitutable powders)

Test procedure	Registration acceptance criteria	Internal acceptance criteria	Comments
Assay	95.0–105.0 % of label	95.0–105.0 % of label	Shelf life limit may be broader, with supporting data. Justification of a broader time of release limit may also be possible. A limit of 90.0–110.0 % of label is usually acceptable for the U.S. A limit of 95.0–105.5 % of label is expected for the EU. For a chiral drug, an enantioselective determination may be needed. For Japan, a microbiological assay is needed for antibiotics
Description or appearance	Descriptive text	Descriptive text	Some countries require a very detailed description. If color changes during storage (i.e., an unavoidable and acceptable change in appearance), a quantitative procedure may be needed. For a solution, Ph.Eur. Degree of coloration as well as clarity and opalescence tests and limits may also be considered. Organoleptic tests should be avoided
Identification			Identification A and B may be IR and HPLC, respectively. Other structure specific methods may be utilized instead of IR. HPLC may be used for lot release. For a chiral drug, a chiral specific method may be needed
A	Positive Infrared adsorption spectrum conforms to that of standard	Positive Infrared adsorption spectrum conforms to that of standard	
B	Positive Retention time of major sample peak corresponds to that of standard	Positive Retention time of major sample peak corresponds to that of standard	
Impurities			Safety information is needed for the qualification of specified impurities above threshold limits. Qualification thresholds are different for human and veterinary products. For a chiral drug, control of the other enantiomer is necessary if it is a degradation product
Total	Limit supported by batch data and stability data	Limit supported by batch data and stability data	
Specified degradation products	Limit based on the qualification level and supported by batch data and stability data	Limit based on the qualification level and supported by batch data and stability data	
Unspecified degradation products	Human health: Limited based on maximum daily dose and ICH threshold limits	Human health: Limited based on maximum daily dose and ICH threshold limits	

(continued)

Table 3.22 (continued)

Test procedure	Registration acceptance criteria	Internal acceptance criteria	Comments
Dose uniformity	Meets USP, Ph.Eur., and JP requirements	Meets USP, Ph.Eur., and JP requirements	Evaluate global compendia requirements for methods and limits. Compendia fill requirements must also be considered (minimum fill and deliverable volume)
pH	Limit supported by batch data and stability data	Limit supported by batch data and stability data	Consider, e.g., stability potency/impurities, solubility and in-process control limits
Microbial quality	Limit based on compendia requirements	Limit based on compendia requirements	Micro count/pathogens. Evaluate global compendia requirements for methods and limits
Preservative content	Limit support by batch data, stability data, and AET data (for antimicrobial preservatives)	Limit support by batch data, stability data, and AET data (for antimicrobial preservatives)	Consider both antimicrobial preservatives antioxidants. A limit of 90.0–110.0 % of label at time of release may be expected; shelf life limit may be broader, with supporting data. For antimicrobial preservatives, preservative effectiveness should be demonstrated during development, scale up, and throughout shelf life[20]. Evaluate global compendia requirements for AET methods and limits. Some markets (e.g., EU) also require a preservative identification test, and HPLC may be adequate. Evaluate global compendia requirements for antioxidant method and limits
Extractables	Limit supported by batch data and stability data	Limit supported by batch data and stability data	Safety information will be needed. Consider both specific and non-specific (global compendia) tests
Alcohol content	Limit consistent with labeling	Limit consistent with labeling	Needed when alcohol content is declared quantitatively on the label. May be assayed or calculated
Dissolution	Limit supported by batch data, stability data and bioavailability data	Limit supported by batch data, stability data and bioavailability data	Evaluate global compendia requirements for acceptance criteria
Particle size distribution	Limit supported by batch data, stability data and dissolution data	Limit supported by batch data, stability data and dissolution data	Russia requires test for particle size in constituted suspension or compendia test for sedimentation stability
Redispensability	Limit supported by batch data, stability data	Limit supported by batch data, stability data	May be needed for a suspension which settles on storage. Particle size is a factor, and assay/homogeneity may be affected

(continued)

Table 3.22 (continued)

Test procedure	Registration acceptance criteria	Internal acceptance criteria	Comments
Rheological properties	Limit supported by batch data and stability data	Limit supported by batch data and stability data	May be needed for relatively viscous drug product
Specific gravity	Limit supported by batch data and stability data	Limit supported by batch data and stability data	
Reconstitution time	Limit supported by batch data and stability data	Limit supported by batch data and stability data	
Water	Limit supported by batch data and stability data	Limit supported by batch data and stability data	A Karl Fischer method is usually described; specify the method, e.g., volumetric or coulometric
Solid state forms	Positive	Positive	Needed when a polymorph or solvate could affect drug product performance, safety, efficacy, bioavailability, or stability–if the form is known to change during manufacture or on stability. A surrogate test is preferred
Excipient identification	Positive	Positive	For the EU, a test for identification of colorants is required. Identification of all excipients has been a requirement for drug products sourced from outside of the EU (test provided to be applied if needed, not included in finished product specifications). Evaluate whether this is needed for each dossier, as the requirement appears to be diminishing

Table 3.23 Topical drug products—investigational phase 1 and 2 (Includes topical creams, ointments, and gels)

Test procedure	Registration acceptance criteria	Internal acceptance criteria	Comments
Assay	95.0–105.0 % of label	95.0–105.0 % of label	Shelf life limit may be broader, with supporting data. Justification of a broader time of release limit may also be possible. A limit of 90.0–110.0 % of label is usually acceptable for the U.S. A limit of 95.0–105.5 % of label is expected for the EU
Description or appearance	Descriptive text	Descriptive text	Some countries require a very detailed description. Organoleptic tests should be avoided
Identification	Positive Retention time of major sample peak corresponds to that of standard	Positive Retention time of major sample peak corresponds to that of standard	HPLC may be utilized. A second identification test is usually not required at this stage
Dose uniformity	Meets compendia requirements	Meets compendia requirements	Evaluate global compendia requirements for methods and limits. Compendia fill requirements must also be considered (minimum fill and deliverable volume)
pH	Report results	Report results	A numeric limit should be considered if pH is stability indicating. In addition to potency/impurities, consider, e.g., solubility and in-process control limits
Microbial quality	–	Limit appropriate to dosage form	Evaluate global compendia requirements for methods and limits
Preservative content	Report results	Report results	Consider both antimicrobial preservatives and antioxidants. For antimicrobial preservatives, preservative effectiveness should be demonstrated during development, scale up, and throughout shelf life[20]. The potential need for a numeric limit is based on evaluation of batch data, stability data, and antimicrobial effectiveness data. Evaluate global compendia requirements for AET methods and limits
Test procedure	Registration acceptance criteria	Internal acceptance criteria	Comments

(continued)

Table 3.23 (continued)

Test procedure	Registration acceptance criteria	Internal acceptance criteria	Comments
Drug release	–	Report results	Internal limit appropriate to dosage form, or data review, may be utilized. Evaluate global compendia requirements for acceptance criteria
Rheological properties		Limit supported by batch data and stability data	
Specific gravity		Report results	
Water		Report results	A Karl Fischer method is usually described; specify the method, e.g., volumetric or coulometric. A numeric limit should be considered if product performance is sensitive to the presence of water
Supplemental specifications (Add when available)			
Impurities	Report results	Report results	During development, while degradation mechanisms are being characterized, it is important to evaluate how process impurities and degradation products will be monitored. The potential need for a numeric limit is based on evaluation of batch data, stability data, and safety
Total degradation products	Report results	Report results	
Individual degradation products			
Impurities Enantiomeric	–	Report results	The potential need for a numeric limit is based on evaluation of batch data, stability data, and safety
Solid state forms	Positive	Positive	Needed when a polymorph or solvate is present. A surrogate test may be utilized

Table 3.24 Topical drug product—phase 3 and marketed product (Includes topical creams, ointments, and gels)

Test procedure	Registration acceptance criteria	Internal acceptance criteria	Comments
Assay	95.0–105.0 % of label	95.0–105.0 % of label	Shelf life limit may be broader, with supporting data. Justification of a broader time of release limit may also be possible. A limit of 90.0–110.0 % of label is usually acceptable for the U.S. A limit of 95.0–105.5 % of label is expected for the EU. For a chiral drug, an enantioselective determination is needed. For the Japan a microbiological assay is needed for antibiotics
Description or appearance	Descriptive text	Descriptive text	Some countries require a very detailed description. If color changes during storage (i.e., an unavoidable and acceptable change in appearance), a quantitative procedure may be needed. Organoleptic tests should be avoided
Identification			Identification A and B may be IR and HPLC, respectively. Other structure specific methods may be utilized instead of IR. HPLC may be used for lot release. For a chiral drug, a chiral specific method may be needed
A	Positive Infrared adsorption spectrum conforms to that of standard	Positive Infrared adsorption spectrum conforms to that of standard	
B	Positive Retention time of major sample peak corresponds to that of standard	Positive Retention time of major sample peak corresponds to that of standard	
Impurities			Safety information is needed for the qualification of specified impurities above threshold limits. Qualification thresholds are different for human and veterinary products. For a chiral drug, control of the older enantiomer is necessary if it is a degradation product
Total	Limit supported by batch data and stability data	Limit supported by batch data and stability data	
Specified degradation products Unspecified degradation products	Limit based on the qualification level and supported by batch data and stability data Human health: Limited based on maximum daily dose and ICH threshold limits	Limit based on the qualification level and supported by batch data and stability data Human health: Limited based on maximum daily dose and ICH threshold limits	

(continued)

Table 3.24 (continued)

Test procedure	Registration acceptance criteria	Internal acceptance criteria	Comments
Dose uniformity	Meets USP, Ph.Eur., and JP requirements	Meets USP, Ph.Eur., and JP requirements	Evaluate global compendia requirements for methods and limits. Compendia fill requirements must also be considered (minimum fill and deliverable volume)
pH	Limit supported by batch data and stability data	Limit supported by batch data and stability data	Consider, e.g., stability potency/impurities, solubility and in-process control limits
Microbial quality	Limit based on compendia requirements	Limit based on compendia requirements	Micro count/pathogens. Evaluate global compendia requirements for methods and limits
Preservative content	Limit support by batch data, stability and AET data (for antimicrobial preservatives)	Limit support by batch data, stability and AET data (for antimicrobial preservatives)	Consider both antimicrobial preservatives antioxidants. A limit of 90.0–110.0 % of label at time of release may be expected; shelf life limit may be broader, with supporting data. For antimicrobial preservatives, preservative effectiveness should be demonstrated during development, scale up, and throughout shelf life[20]. Evaluate global compendia requirements for AET methods and limits. Some markets (e.g., EU) also require a preservative identification test, and HPLC may be adequate. Evaluate global compendia requirements for antioxidant method and limits
Extractables	Limit supported by batch data and stability data	Limit supported by batch data and stability data	Safety information will be needed. Consider both specific and non-specific (global compendia) tests
Alcohol content	Limit consistent with labeling	Limit consistent with labeling	Needed when alcohol content is declared quantitatively on the label. May be assayed or calculated
Drug release	Limit supported by batch data, stability data bioavailability data	Limit supported by batch data, stability data bioavailability data	Evaluate global compendia requirements for acceptance criteria

(continued)

Table 3.24 (continued)

Test procedure	Registration acceptance criteria	Internal acceptance criteria	Comments
Rheological properties	Limit supported by batch data and stability data	Limit supported by batch data and stability data	Related tests, such as droplet size in creams, should be considered
Specific gravity	Limit supported by batch data and stability data	Limit supported by batch data and stability data	
Water	Limit supported by batch data and stability data	Limit supported by batch data and stability data	A Karl Fischer method is usually described; specify the method, e.g., volumetric or coulometric
Solid state forms	Positive	Positive	Needed when a polymorph or solvate could affect drug product performance, safety, efficacy, bioavailability, or stability–if the form is known to change during manufacture or on stability. A surrogate test is preferred
Excipient identification	Positive	Positive	For the EU, a test for identification of colorants is required. Identification of all excipients has been a requirement for drug products sourced from outside of the EU (test provided to be applied if needed, not included in finished product specifications). Evaluate whether this is needed for each dossier, as the requirement appears to be diminishing

- The API in a capsule should be put on an abbreviated stability study, since the API stability program can't support the stability of the API in a capsule (Table 3.13).

Examples of specifications for drug substance and various formulated drug product are described in the Tables 3.14, 3.15, 3.16, 3.17, 3.18, 3.19, 3.20, 3.21, 3.22, 3.23, 3.24.

3.3.2 Stability Management Process [9]

Responsibilities of the Stability Group

- Initiation of Stability Studies
- Writing the Protocol
- Generation of Data
- Review and Reporting Results
- Interpretation of Data
- Writing of Stability Reports

The Goals of a Stability Study

- A stability study is a structured program designed to monitor and test a drug substance or drug product according to its intended use and its own unique specifications, under specified conditions for a pre-determined time period.
- The stability study should provide evidence on how the quality of a drug substance or drug product changes with time under the influence of a variety of environmental factors such as:
  - Temperature, Humidity, and Light.
- The intended use of the study
  - Recommend storage conditions.
  - Establish a retest period for the drug substance.
  - Determine shelf life for the drug product.

Prior to the design of the Stability Studies, Force Degradation studies should be implemented and the data taken into consideration for the stability study design. The following information can be used to implement force degradation studies:

- Conditions of degradation are dependent on the material being tested:
  - Bulk Chemical
  - Solid Dosage Forms
  - Liquid/IV Dosage Forms
  - Solutions/Suspensions

Table 3.25 Conditions for force degradation of various materials

Conditions	Chemical	Solid dosage forms	Liquid/IV/ susp. solutions
High temp- solid	10 °C below DSC melt	75–120 °C-tablets NMT 75 °C –cap content. NMT 50 °C- capsules	–
Mid temp- solid	10–20 °C below high Temp	50–60 °C	–
Low temp- Solid	55–75 °C	40 °C	
45 °C/75 %RH Open container	+	+	–
ICH photo conditions	+	+	+
High temp-solution (NMT 90 °C)	+	–	+
Mid temp-solution (10–20 °C below high temp)	+	–	+
Low temp-solution (40–55 °C)	+	–	+
Acidic solution (0.1 N HCl)	+	–	–
Basic solution (0.1 N NaOH)	+	–	–
Oxidized (Peroxide)	+	–	–

- The Table 3.25 contains a comprehensive list of condition to consider. The conditions chosen should be included in the force degradation protocol

The amount of degradation should be limited to the range of 5–15 % (75 % of the parent compound remaining); so that we are limiting degradation to the formation of mostly primary degradation products and we have a good probability of achieving reasonable mass balance.
Elements of a Stability Protocol

- Purpose
- Compliance Statement
- Stability Study Number
- Protocol Number
- Drug Substance (API)
  - Lot Number
- Formulation Description
  - Lot Numbers
- Container Closure system
  - Package Lot Numbers

- Experimental
  - Equipment

    Stability Chambers
    Data Acquisition System
    Instrumentation Information
    Statistical Program

  - Time Points
  - Storage Conditions
  - Testing Plan
- Description of Test at each Time Point and Condition
- Label Information
- Reference Standard

Elements of the Stability Report

- Product Name
  - Lot Numbers
- API Lot Number
- Stability Study Number
- Protocol Number
- Package Lot numbers
- Packager, Name, Location
- Package Date
- Container Closure Description
- Manufacturer, Name, location
- Manufacture Date
- Study Initiation date
- Storage Conditions
- Time points
- Data Table (for each condition)
  - Specifications
  - Condition
  - Analytical methods/Tests
  - Time Points
  - Data for each Specification
- Conclusions
- Statistical Analysis of the Data (if necessary)
- Shelf Life or Re-Testing Dating
- Supportive Stability Data
  - Force Degradation Studies
  - Development Stability Studies

- Photo Stability in Commercial packaging configuration (Table 3.26)

Peaks that are formed during stability studies must be qualified as degradation products to be included in the stability Study (Fig. 3.9).

Statistical Evaluation of Stability Data

- *There are Several statistical approaches to Stability Data Analysis*:
  - *Linear Regression*
  - *Poolability Testing*
  - *Statistical Modeling*
- Data Evaluation for RE-Test or Shelf Life at Room Temperature
  - No significant Change at Accelerated and Long term conditions, statistical analysis normally considered unnecessary but justification needs to be provide

    Relevance of accelerated Data
    Mass Balance
    Other Supporting Data

  - Extrapolation of Re-Test or Shelf Life Dating can be Twice the long Term period, but can not exceed 12 months beyond the long term period
- Significant Change/Variability at Accelerated or Long term conditions, Statistical analysis of the Long Term Data is recommended If Significant Change/ Variability is observed in the accelerated condition, re-test or shelf life dating will depend on the data from the intermediate and long term conditions
  - When differences in stability between batches are observed and the data is not poolable than the re-test or shelf life period cannot exceed the shortest period supported by any batch.
- Significant Change/Variability at Accelerated or Long term conditions, Statistical analysis of the Long Term Data is recommended
  - Extrapolation beyond the period covered by long-term data can be proposed; however, the extent of extrapolation would depend on whether long-term data are amenable to statistical analysis.
  - Data not amenable to statistical analysis

    Extrapolated re-test or shelf life can be 1.5 times the long term condition period, but not exceeding 6 months beyond the long term period

  - Data amenable to statistical analysis

    Extrapolation of Re-Test or Shelf Life Dating can be Twice the long Term period, but cannot exceed 12 months beyond the long term period
    Where significant change/variability at the intermediate condition, the proposed re-test or shelf life dating should not exceed the long term period

Table 3.26 Example of data reporting for 25 °C/60 %RH condition

		Time points (Month)		
Test	Specifications	Initial	2 months	3 months
Appearance	Clear, colorless liquid	Conforms	Conforms	Conforms
Identification (HPLC)	Retention time of the main peak matches the standard (±2 %)	Conforms	Conforms	Conforms
Assay (%LC)	90–110 %	*Prep*	*Prep*	*Prep*
		1 99.4	1 96.7	1 99.3
		2 98.7	2 96.3	2 99.4
		3 99.2	3 97.3	3 100.2
		Mean = 99.2	Mean = 96.8	Mean = 99.6
		%RSD = 0.1	%RSD = 1.5	%RSD = 0.5
Degradation products (HPLC) (% Area)	NMT 2 %	Mean = 1.4 %	Mean = 1.4 %	Mean = 1.6 %
Specified (name/RRT)	NMT 2 %	%RSD = 5.2	%RSD = 5.1	%RSD = 8.2
Unspecified (RRT)	NMT 3 %	Mean = 0.75 %	Mean = 0.60 %	Mean = 0.9 %
Total		%RSD = 3.0 %	%RSD = 5.2 %	%RSD = 10.3 %
		2.15 %	2.0 %	2.5 %
Anti-Oxidant	80–120%	*Prep*	*Prep*	*Prep*
		1 94.9 %	1 92.9 %	1 93.1 %
		2 90.5 %	2 93.4 %	2 91.7 %
		3 93.5 %	3 92.4 %	3 93.7 %
		Average = 93.0 %	Average = 92.8 %	Average = 92.9 %
		%RSD = 2.4	%RSD = 1.7	%RSD = 1.)
Microbial enumeration	NMT 2000 CFU/g	<10 CFU/g	Not Determined	Not determined
Total aerobic count	NMT 200 CFU/g	<10 CFU/g		
Total yeast and molds				

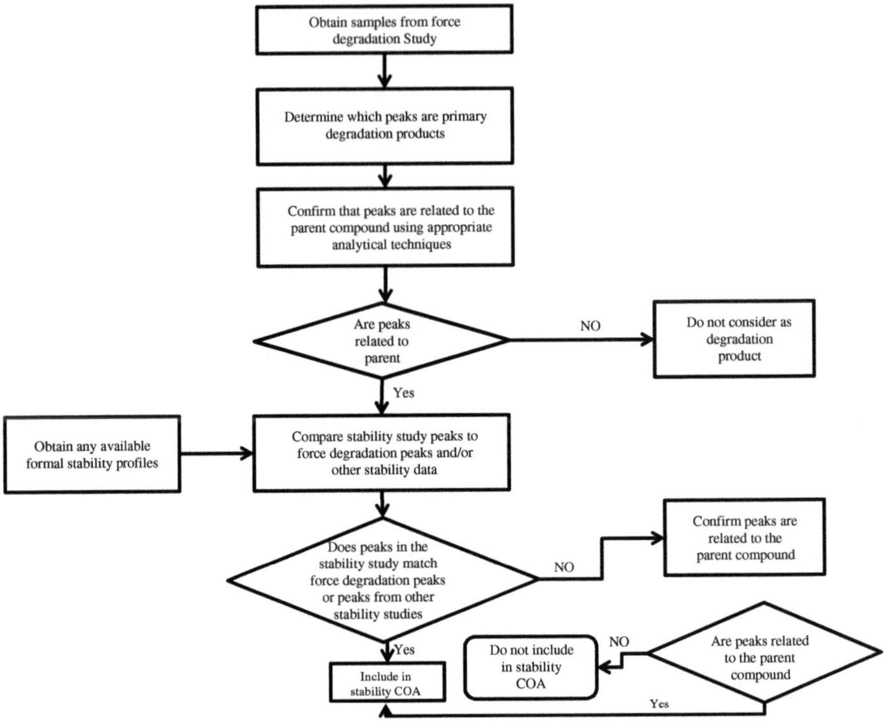

Fig. 3.9 Qualification of chromatographic peaks as degradation products

- Data Evaluation for RE-Test or Shelf Life below Room Temperature (Refrigerator)
  - No significant Change at Accelerated and Long term conditions, statistical analysis normally considered unnecessary but justification needs to be provide

    Relevance of accelerated Data
    Mass Balance
    Other Supporting Data
    Extrapolation of Re-Test or Shelf Life Dating can be 1.5 times the long Term period, but cannot exceed 6 months beyond the long term period

- Significant Change/Variability at Accelerated or Long term conditions, Statistical analysis of the Long Term Data is recommended

  If Significant Change/Variability is observed in the accelerated condition, re-test or shelf life dating will depend on the data from the long term conditions. When differences in stability between batches are observed and the data is not poolable than the re-test or shelf life period can not exceed the shortest period supported by any batch.

- Extrapolation beyond the period covered by long-term data can be proposed; however, the extent of extrapolation would depend on whether long-term data are amenable to statistical analysis.

Data not amenable to statistical analysis

- Extrapolated re-test or shelf life at the long term condition period, but not exceeding 3 months beyond the long term period

Data amenable to statistical analysis

- Extrapolation of Re-Test or Shelf Life Dating can be 1.5 times the long Term period, but can not exceed 6 months beyond the long term period
- Where significant change/variability at the accelerated condition, 3 and 6 months is observed the proposed re-test or shelf life dating should not exceed the long term period
- Where significant change/variability at the accelerated condition at 3 months is observed re-test and shelf life dating at the long term period, should be confirmed by statistical analysis and discussion should provide as to excursions outside the labeled storage condition
- Data Evaluation for RE-Test or Shelf Life below Room Temperature (Freezer)
  - For drug substances or products intended for storage in a freezer, the retest period or shelf life should be based on long-term data.
  - In the absence of an accelerated storage condition, testing on a single batch at an elevated temperature should be conducted to address the effect of short-term excursions outside the proposed label storage condition
  - For drug substances or products intended for storage below −20 °C, the retest period or shelf life should be based on long-term data and should be assessed on a case-by-case basis.

Examples of Statistical Approaches to Stability Data Analysis [10]

- Linear regression, poolability tests, and statistical modeling, are examples of statistical methods and procedures that can be used in the analysis of stability data for which there is a proposed acceptance criterion.
  - In general, the relationship between certain quantitative attributes and time is assumed to be linear.
  - The regression line along with its 95 % Confidence intervals is compared with upper and lower acceptance criteria.
  - The time where the 95 % confidence intervals intersects the acceptance criteria, is the time period which can be supported. Extrapolation of the time period can be applied as per the discussions above.
  - This approach can be utilized to estimate the re-test or shelf life period for a single batch, or individual batches. When applying this approach to the combination of multiple batches, appropriate statistical test must be performed first.

- For a drug substance or for a drug product the retest period or shelf life is generally estimated based on the stability data from a minimum of three batches. When analyzing data, two statistical approaches can be considered.
  - The objective of the first approach is to determine whether the data from all batches support the proposed retest period or shelf life.
  - The objective of the second approach, testing for poolability, is to determine whether the data from different batches can be combined for an overall estimate of a single retest period or shelf life.
  - If all shelf lives estimated by the original model are longer than the proposed shelf life, further model building is considered unnecessary and the proposed shelf life will generally be appropriate
  - If one or more of the estimated shelf lives fall short of the proposed shelf life, model building can be employed
- Analysis of covariance can be employed to test the difference in slopes and intercepts of the regression lines among factors and factor combinations. The purpose of the procedure is to determine whether data from multiple factor combinations can be combined for the estimation of a single shelf life.
  - The tests for poolability should be specified to determine whether there are statistically significant differences among factors and factor combinations.
  - All tests should be conducted using appropriate levels of significance. It is recommended that a significance level of 0.25 be used for batch-related terms.
  - If the tests for poolability show that the data from certain factors or factor combinations should not be combined, a single shelf life can be estimated based on the shortest estimated shelf life among all levels of factors and factor combination.

Glossary

- *Stress Testing*. Involves exposing a drug product or drug substance to conditions (temperature and humidity) that induce the formation of degradation products. In a stability program stress testing enables one to approximate the amount of degradation one could expect under typical storage conditions of the drug product or drug substance at the future time points.
- *Impurity*. Any component of the new drug substance that is not the chemical entity defined as the new drug substance. Typically, an impurity is not a decomposition product of the drug product although it may be a decomposition product of the drug substance.
- *Specified Impurity*. An impurity that is individually listed and limited with a specific acceptance criterion in the new drug substance specification. A specified impurity can be either identified or unidentified.
- *Unspecified impurity*. An impurity that is limited by a general acceptance criterion, but not individually listed with its own specific acceptance criterion, in the new drug substance specification.

- *Degradation Product.* Any component of the new drug substance or drug product that is not the chemical entity defined as the new drug substance, is formed by a chemical reaction from the drug substance itself or by a reaction of the drug substance with other components in the sample, including components from the formulation and/or the packaging used to store the drug product/drug substance.
- *Specified Degradation Product.* A degradant that is individually listed and limited with a specific acceptance criterion in the new drug substance or drug product specification. A specified degradant can be either identified or unidentified.
- *Unspecified Degradant.* A degradant that is limited by a general acceptance criterion, but not individually listed with its own specific acceptance criterion, in the specification for the new drug substance or formulation.

3.3.3 Reference Standard Certification Process [11]

This Process describes the submission, evaluation, certification, distribution and re-certification of Analytical Reference Standards.

- Procedure for reference standard Certification
  - Request for Analytical Reference Standard CertificationReference Standard Certification

    The Requester submits the material and the first page of the certification request form to the Reference Standard Manager (RSM). Based on the certification class of reference standard the Analytical Lead (AL) identifies the required tests and submits the second page of the form to RSM. Based on the availability of the compound, the AL may alter the amounts and required tests.

  - Registration of Material into Reference Standard Program

    The RSM logs material in the reference standard logbook recording all available information regarding the lot, assigns a unique analytical reference standard number (AS number; AS000XXX). AS numbers are assigned in a sequential order. The RSM creates an AS folder using the newly assigned number as the title.

  - Reference standard material in multiple containers

    When multiple containers of the same reference standard are received, each individual container must be labeled such that each container is distinguishable. The first container will be certified as per acceptance criteria for certification; subsequent containers will be certified as per acceptance criteria for re-certification plus identification test.

– Submission of Material for Analysis

The Reference Standard Manager (RSM) weighs an aliquot of the material for testing, records weighing amount on Certification Request form and in log-book, submits material for analysis, either in-house or outsourced, and places the material in quarantine (Fig. 3.10)

Page 1

Requested By:________________________	Date Needed:__________
Department:__________________________	Phone#______________
Certification:___________________	Re-Certification:________
Requester Signature:__________________	Date:________________

Analytical Reference Standard Class

___ Class I ____Class II ____Class III ____Class IV

ANALYTICAL REFERENCE STANDARD REQUEST

Reference Standard Number (AS#):______________________________

Company Compound Number:______________________________

Description *:______________________________

Salt Form *:______________________________

Molecular Weight *:______________________________

Amount of Compound *:______________________________

Manufacturer *:______________________________

Manufacturer Lot Number *:______________________________

Date of Manufacture *:______________________________

Chemist *:______________________________

Notebook Reference *:______________________________

Recommended Storage Condition *:______________________________

Special Instructions:______________________________

Completed by RSM

Date Dispensed:______________________________

By:______________________________

Actual amount weighed:______________________________

Recorded in logbook:______________________________

(* These fields are not required for re-certifications)

Fig. 3.10 Analytical reference standard certification (or Re-Certification) request form page 1

Page 2

Test	Test Method (e. g. QCT, USP)	*Amount Needed for Each Replicate	Number of Replicates	Additional Comments
Appearance		0 mg		
Enantiomeric Purity (Chiral Compounds)		5 mg		
GC Residual Solvent		20mg		Solvents to be tested for:
GC Purity		30 mg		
HPLC Purity		5 mg		
Ion Chromatography		20 mg		Ions to be tested for:
IR Spectroscopy		5 mg		
Karl Fischer (Water Content)		20 mg		Estimated Water Content:
Loss on Drying		1 g		
Heavy Metals		250 mg		
Mass Spectrometry		1 mg		
NMR 1H		5 mg		
NMR 13C		50 mg		
Isotopic Purity		5 mg		
pH		5 ml		
Residue on Ignition		1 g		
DSC		10 mg		
TGA		10 mg		
UV Spectroscopy		5 mg		
X-Ray Diffraction		20 mg		
Additional Test				

*The amounts listed for testing are served as a guideline. They may be altered by the Analytical Lead to accommodate specific assay requirements.
Comments:
Analytical Lead signature: ____________________ Date:______________

Fig. 3.10 (continued)

– Sample Analysis and Data Review

Testing of samples and reporting of data should follow instructions described in the analytical method. For purity analysis (chromatographic), the conditions from the method and system suitability testing of the tested compound are followed, however reference standards are not required since purity is determined as percent of main peak area/total peak area (%a/a). The exception is when Class I reference standards are used to certify other standards. The number of replicates and the sample size for each test are determined by the AL. The results of the analysis are recorded on the second page of the request form. Special instructions and deviations from the analytical method along with justifications are also noted on the second page of the request form. The analyst reports results in laboratory notebook and Report of Analysis (RoA) and submits data package to the AL/RSM. The AL/RSM reviews data and RoA. The analyst submits copy of reviewed RoA to the RSM.

– Retest Date Determination

  The RSM and the AL determines whether there is an equivalent lot to the reference standard lot on stability, or an equivalent lot used previously as reference standard lot. If there is, stability data or re-certification data of an equivalent lot, it may be used to establish retest date for the reference standard lot. If there is no historical data, a default retest date will be assigned. As a default, Class I, Class II and Class IV reference standards will be evaluated every six (6) months; and Class III every twelve (12) months to determine continued suitability for use.

– Certificate of Analysis (CoA) Generation

  The RSM generates the CoA and submits the complete data package to the Analytical Department and the Quality Assurance for approval. The RSM distributes the approved CoA to reference standard users.

– Material Transfer and Documentation

  The RSM transfers the approved reference standard material from quarantine to the active reference standard location. The RSM also updates all current files and stores all related documents.

- Procedure for Re-Certification of Analytical Standards

  – Re-Certification Evaluation

    Prior to the expiration of the reference standard certification period the RSM notifies all users and requests whether the reference standard should be re-certified.

    If re-certification is required, RSM reviews historical (re-certification or stability) data and determines whether retesting is required. If historical data indicates that no significant changes are occurring then retesting is not required and the retesting period may be extended. The historical data must consist of at least three data points to extend the retesting period, however the retesting period cannot be greater than 24 months. The RSM issues a justification memo, generates a new CoA extending the retest date and submits the data and documents to the Analytical Department and Quality Assurance for approval.

  – Re-Certification Testing

    If retesting is required, RSM asks users to submit Recertification Request form. The RSM weighs an aliquot of the material for testing, records and documents weighing amount and submits material for analysis.

– Termination of a Reference standard lot

The reference standard lot may be terminated due to project cancellation or depletion of material. The RSM documents the reason for termination and transfers material to the expired reference standard location. After a reference standard lot has expired, if there is a need to bring the material back to the reference standard program full certification testing will be required as a new lot.

- Inventory Control, Distribution and Tracking of Reference Standards

– Inventory Control

All certified reference standard material is stored in specified temperature conditions and locked chambers. Only the RSM, or designee, has access to the chambers. All reference standard material should be obtained from the RSM, or designee.

– Request for Distribution of Reference Standard

When needed, the requester submits an reference standard request form to the RSM. The material may be used in-house or shipped too external locations (Fig. 3.11).

REQUESTER
Requested By:__________ Date Needed:______
Department:__________ Phone #______
Approver signature:__________ Date:______

ANALYTICAL REFERENCE STANDARD REQUEST
Reference Standard Number (AS#):__________
Compound Number:__________
Quantity Needed:__________

PURPOSE:__________

DELIVER TO (within Company) **OR** **SHIP TO (outside)**
Name:__________
Department:__________
Location:__________

Name of Institution:__________
Attach: Approved Shipping Request Form
Certificate of Analysis (CoA)
Material Safety Data Sheet (MSDS)

REFERENCE STANDARD MANAGER
Dispense of Material
Date Dispensed:__________ Effective Date:______
By:__________ Retest Date:______
Actual amount weighed:__________ Logbook page No.#:______

Receiver
Received by:__________ Department:__________ Date:______

Fig. 3.11 Analytical reference standard request form

- Distribution of Reference Standard

  The RSM, or designee, weighs the reference standard material according to the request and labels the container. The label should include, but is not limited to, the Reference Standard number, Class, and storage condition. The RSM also documents the amount on the request. If the reference standard is to be used in-house, the RSM delivers the material to the requester. The requester signs for the material upon receipt of the material. The requester is to check the current CoA for retest date and purity of the material before each use. If the reference standard is to be shipped to an external facility, the requester submits a Shipping Request, along with the current CoA and MSDS to the RSM. The RSM then delivers the material and the above documents to Shipping Department personnel. The Shipping personnel signs for the reference standard material and documents upon receipt of the material, and returns to the RSM the complete Shipping Request and the receipt of delivery from the shipping company (e.g., FedEx). The RSM files the Reference Standard Request, the Shipping Request, and the receipt of delivery from the shipping company.

- Inventory Tracking

  The RSM, or designee, documents the amount of reference standard removed, the requester's name, purpose of reference standard request, the date and the remaining material balance. Periodically the RSM visually checks all reference standards and alerts users when reference standard supply is getting low.

- Issued Reference Standard

  It is the responsibility of the recipients to ensure that Reference Standards issued are stored and handled under recommended storage condition to maintain their validity. Even if the issued reference standard is stored as directed on the label, the extension of the retest date will not apply to the originally issued material upon re-certification of the reference standard.

• Classification of Reference Standards

  - Reference standards are classified into four classes as shown in Table 3.27

• Class I Primary Active Pharmaceutical Ingredient (API) Reference Standard

  - Class I standards should be used sparingly and should only be used for purposes other than certifying other standards when Class II standards are not available.
  - Identity Specification
    Structural ID—Spectral by two independent methods (e.g., HPLC, NMR, MS or IR).
  - Specification
    Class I reference standard lots must meet the specifications shown in Table 3.28.

Table 3.27 Classification of reference standards[a]

Class	Description	Intended uses	Purity by HPLC (%a/a) requirement
1	Primary active pharmaceutical ingredient (API)	To certify other standards for quantitative analysis	≥99 %
2	Secondary API	For quantitative analysis	≥97 % for phase II ≥98 % for phase III
3	Qualitative standard	For qualitative analysis: internal standard, system suitability, marker, racemate, stable isotope (deuterated)	Dependent on the intended use
4	Non-API quantitative standard: Metabolite, impurity, degradant, starting material, intermediate, etc.	For quantitative analysis of impurities, degradants, bioanalytical assays	≥90 %

[a] API materials of biological/biotechnological, peptide, oligonucleotide, radiopharmaceutical, fermentation product and semi-synthetic products derived therefrom, herbal products, and crude products of animal or plant origin are exceptions due to intrinsic properties of the molecule, they do not need to meet the acceptance criteria in the table

Table 3.28 Class I test and re-test scheme for characterization

Required tests	Acceptance criteria	Required re-tests	Acceptance criteria
Enantiomeric purity (Chiral compounds only)	≤0.5 % of the undesirable enantiomer	Enantiomeric purity (Chiral compounds only)	≤0.5 % of the undesirable enantiomer
Residual solvent by GC	≤0.5 % of each individual solvent, unless solvate	NR	NR
Purity by HPLC	≥99.0 % purity	Purity by HPLC	≥99.0 % purity
Counter ion(s)	Report results—consistent with salt stoichiometry	NR	NR
Identity	Must conform to structural identity, via two methods	NR	NR
Water content	≤0.5 %, unless hydrate	Water content	≤0.5 %, unless hydrate
Residue on ignition	≤0.5 %	NR	NR
Appearance	Material specific	Appearance	No significant change from original material description

NR Not required

- Class II API Quantitative Reference Standard
  - Purity for Class II reference standard by HPLC follows a phase based system as shown in Tables 3.29, 3.30.
- Class III Qualitative Reference Standards
- Class III Reference standards lots must meet the criteria shown in Table 3.31

Table 3.29 Phase based purity by HPLC specification for class II API

Reference standards	
If in clinical phase I	Highest attainable purity
If in clinical phase II	≥97.0 % purity
If in clinical phase III	≥98.0 % purity
Exceptions to the above: purity for biological/fermentation materials (proteins, antibiotics, etc.)	Due to intrinsic properties of molecule, highest purity obtainable and is representative for this material
Exceptions to the above: solutions	Target concentration prepared from solid RS material

Table 3.30 Class II test and re-test scheme for characterization

Required tests	Acceptance criteria	Required re-tests	Acceptance criteria
Residue on ignition (ROI)	≤1.0 %	NR	NR
Enantiomeric purity (Chiral compounds only)	≤2.0 % of the undesirable enantiomer	Enantiomeric purity (Chiral compounds only)	≤2.0 % of the undesirable enantiomer
GC residual solvent	≤1.0 % of each individual solvent, unless solvate	NR	NR
Purity by HPLC	See Table 3	Purity by HPLC	See Table 3
Counter ion(s)	Report result—consistent with salt stoichiometry	NR	NR
Identity	Must conform to structural identity, via two methods.	NR	NR
Water content	≤2.0 %, unless hydrate	Water content	≤2.0 %, unless hydrate
Appearance	Material specific	Appearance	No significant change from original material description

NR Not required

Table 3.31 Class III Test and Re-Test Scheme for Characterization

Required tests	Acceptance criteria	Required re-tests	Acceptance criteria
[a]Identity	Conforms to identity by one method	NR	NR
[b]Qualitative examination	Material specific	Qualitative examination	No significant change
Appearance	Material specific	Appearance	No significant change
Isotopic purity (for bioanalytical deuterated compound only)	Highest attainable purity[c]	Isotopic purity (for bioanalytical deuterated compound only)	Highest attainable purity[c]

NR Not required

[a] A single method either structural (e.g., UV) or physical (e.g., HPLC) is acceptable

[b] An appropriate qualitative test will be defined for the specific molecule (e.g., HPLC purity)

[c] Isotropic purity should be at a minimum of 99.0 %; however more than 99.5 % is desirable

Table 3.32 Class IV test and re-test scheme for characterization

Required tests	Acceptance criteria	Required re-tests	Acceptance criteria
Residue on ignition (ROI)	Report results	NR	NR
Entantiomeric purity (Chiral compounds only)	≤5.0 % of the undesirable enantiomer	Entantiomeric purity (Chiral compounds only)	≤5.0 % of the undesirable enantiomer
GC residual solvent	≤3.0 % of each individual solvent, unless a solvate	NR	NR
Purity by HPLC	≥90 % purity	Purity by HPLC	≥90 % purity
Counter ion(s)	Report result—consistent with salt stoichiometry	NR	NR
Identity	Must conform to structural identity, via one method.	NR	NR
Water content	Report results	Water content	No significant change
Appearance	Material specific	Appearance	No significant change form original material description

NR Not required

- Class IV non- API Quantitative Reference Standard
- Class IV non API (e.g., Impurities, degradation products, intermediates, starting materials, etc.) reference standard lots should meet the criteria shown in Table 3.32

3.3.4 Training [12]

FDA-regulated companies are required to document what training they administer to their employees, these companies must also keep all the training documentation current to when all training is completed or updated. This requirement by the FDA was imposed to control all products that can affect consumers, safety and quality of life, such as food and prescription medications.

Training is required by all personnel that will be involved in cGMP, GLP, and GCP activities. Each staff member is required have at a minimum the following training.

Orientation/Safety Training. Training begins on the first day of employment and is designed to introduce employees to company policies and procedures. Training includes coverage of general safety practices including emergency information, instruction on reporting work-related accidents, and an introduction to the safety manual.

Documentation Training. Personnel are required to 'read and understand' applicable documents. Documented evidence of this review of documents is to be included in training files. This review provides assurance that the employee has a basic understanding of job requirements and methods. Further technical training and on-the-job training allow for skill attainment.

Regulatory Compliance Training. Training in the applicable regulatory requirements (e.g., Good Manufacturing Practice Regulations, Good Laboratory Practices Regulations, and/or Good Clinical Practice Regulations) is offered to appropriate employees. In addition, employees are provided with appropriate regulatory review/update training programs.

Specialized Training. Personnel may receive specialized individual or group training programs as appropriate. All members of the analytical staff must have documented training on all analytical technology used for GMP analyses. Examples of the type of information included in the training modules for the Analytical Department Staff are described below:

- Basic Analytical Skills
  - Glassware (usage and storage)

    Volumetric
    Graduated

  - Filters
  - Pipetting
  - Syringes
  - Melting Point

    USP
    DSC

  - Specific Gravity
  - Thermometers
  - Dispensing Liquids
  - Waste Disposal
  - Basic Analytical Calculations
- Laboratory Notebooks documentation and Storage
  - Issuance
  - Language
  - Table of Contents
  - Glossary
  - Corrections
  - Cross Reference
- Reporting of Analytical Data
  - Units
  - Rounding
  - Significant figures
  - Sample Replicates

- Review and Approval of Data
  - Raw Data
  - GMP/GLP Compliance
- Electronic Balances
  - Instrument Setup

    Level Balance
    Zero Balance
- Calibration
  - Internal
  - Using Weights

    Basic Operation
    Maintenance
- Use of pH Meters
  - Basic Theory
  - Electrodes

    Reference
    Glass
    Combination
    Storage
  - Buffers

Calculation/Preparation

  - Calibration Procedure
  - Basic Operation
  - Routine Maintenance
- Titration
  - Aqueous
  - Non- Aqueous
  - Karl Fischer
  - Oxidation–Reduction
  - Complex metric
- Gas Chromatography (GC)
  - Basic Theory
  - Instrument Setup

IQ, OQ, PQ

  - Calibration Procedure
  - Instrument Operation
  - Data Acquisition System

- Sample Preparation

  - Liquid
  - Head Space
  - Solid

- Column Chromatography Phases

  - Liquid
  - Solid

- Detection Modes

  - FID, NPD, EC
  - Tandem (MS, IR, etc.)

- Trouble Shooting and Routine Maintenance
- High Performance Liquid Chromatography (HPLC)

  - Basic Theory
  - Instrument Setup

IQ, OQ, PQ

  - Calibration Procedure
  - Instrument Operation
  - Data Acquisition System
  - Sample Preparation
  - Column Chromatography Phases

    Reverse Phase
    Normal Phase
    Ion Exchange
    Size Exclusion
    Gel Permeation
    Vacancy Chromatography
    Chiral
    Affinity

  - Column Care and Storage
  - Solvent Strength and Selectivity
  - Detection Modes

    UV/Visable
    Refractive Index
    Evaporative Light Scattering
    Tandem (MS, IR, NMR, etc.)

  - Trouble Shooting and Routine Maintenance

- Dissolution
  - Basic Theory

USP (Current Version)

- Instrument Setup
  - IQ, OQ, PQ
- Calibration Procedure
- Instrument Operation
- Data Acquisition System
- Sample Preparation
- Mixing Device
  - Paddle
  - Basket
  - Flow through
- Media
  - Aqueous
  - Buffers pH Ranges
  - Additives

    Enzymes
    Surfactants
    Co-Solvents

  - Sync Conditions
- InVitro–InVivo Correlations (IVIVC)
  - USP Correlation Classes (I, II, III)
  - Biological Classification System (BCS)
- Detection Modes
  - UV/Visible
  - HPLC
- Infra-Red (IR)/Fourier Transform IR Spectroscopy
  - Basic Theory
  - Instrument Setup

IQ, OQ, PQ

- Calibration Procedure
- Instrument Operation
- Data Acquisition System
- Sample Preparation

- KBr Pellet
- NaCl Cells
- Attenuated Total Reflectance
- Nujol Mull
- Diffuse Reflectance Infrared Fourier Transform (DRIFTS)

• Light Source

- Nernst Glower

• Detectors

- Quantum
- Thermal

• Ultra Violet/Visable Spectroscopy

- Basic Theory
- Instrument Setup

IQ, OQ, PQ

- Calibration Procedure
- Instrument Operation
- Data Acquisition System
- Sample Preparation

Quartz Cells

- Light Source

  Neon Arc
  Tungsten Lamp

- Monochromators

  Filters
  Prism
  Grating

- Detectors

  Photomultiplier Tube
  Diode Array

• Other Technologies

- Thin Layer Chromatography (TLC)
- Super Critical Fluid Chromatography (SFC)
- Capillary Electrophoresis (CE)

Trainer Qualification. Individuals are considered qualified to perform training after demonstrating competence in the particular technology. The trainer can also be qualified by an outside expert. Training needs to be documented in their training files.

Grandfathering. Personnel employed prior to the effective date of the Training Program may be considered qualified or "grandfathered" on techniques based on previous experience or training. This must be documented and in their training file. "Grandfathering" cannot occur once the Training Program is effective. In the future, all newly hired individuals must go through training irrespective of education and experience.

Training Procedure. The training Procedure consists of three Phases.

- Phase I self-study
  - The trainer will supply the study material, including training module, user manuals, instrument related SOP, and analytical testing methods or other material to the trainee.
  - The trainee shall then read and understand the contents of the assigned documents. The trainee will initial and date once he/she completes the study material.
- Phase II On-The-Job Training

Company Name	Training Module #	Rev #
Technique:		Effective Date
Objective :		
Additional Safeguards Beyond Normal Laboratory Procedures:		

1	General operating principles and theory
2	Location of instruments, operators Guidelines, Manuals, Logbooks
3	General instrument operation
4	Additional module training topics
5	Trouble shooting and Maintenance
6	Training experiments

Trainee Signature:____________________________ Date:____________

Trainer Signature:____________________________ Date:____________

Approved by:	Date:

Fig. 3.12 Example of an analytical training module template

- The Trainer will supply the appropriate training module to the trainee. The training module should provide the references to the general operating principles and theory of the technique.
- The module should include the detailed description of one or several training experiments for the trainee to perform.
- The locations of the instrument, manuals, equipment logbook and calibration/ maintenance log book, and any other relevant references should be discussed with the trainee.
- The trainer should discuss with and demonstrate to the trainee on the general operation of the technique, such as the software, sample preparation, data analysis, and any common trouble shooting procedures.
- The trainer should discuss with the trainee all elements documented in the training module that is required to properly execute the analytical technique.

- Phase III Training Files and Review
  - It is the responsibility of the trainer to forward, in a timely manner, any training forms and the signed training module to the QA department for filing in the training record file of the analytical department trainee.
  - If at any time or during the periodic review of the analyst training record, it is determined that the analyst has not applied the technology for a period of 2 years or longer the analyst must be retrained prior to using the technology for GMP analyses, the completion of the retraining will be recorded (Fig. 3.12).

References

1. Catalano T, Madsen G, Demarest C (2000) HPLC method development. Am Pharm Rev 3:62
2. Faulstich R, Catalano T (1991) Interactive computer optimization of high performance liquid chromatographic separations in pharmaceutical analysis. LC/GC Magazine 9(11):776
3. Catalano T (1991) Pharmaceutical HPLC solvent optimization utilizing Hewlett- Packard 's ICOS software. Monsanto intercompany analytical symposium—TCMTECH, St Louis
4. Snyder LR, Kirkland JJ, Glajch JL (1997) Practical HPLC method development 2nd edn. Wiley, New York
5. Vogel F, Galushko S (20013) Application of chromsword software for automatic HPLC method development and robustness studies. Separation of terbinafine and impurities
6. Dolan JW (2008) The perfect method, part 7: the gradient shortcut, LCGC Eur 21(3)
7. Catalano T (2000) DRUG substance & pharmaceutical dosage form development and validations for global registration meeting, Philadelphia
8. ICH 2Q(R1), Validation of analytical methods
9. Carstensen YT (1990) Drug stability. Marcel Dekker, New York
10. Catalano T (1999) International stability programs meeting, Philadelphia
11. Browne DC (2009) Reference standard material qualification. Pharm Technol 33(4):66–73
12. 21CFR 211(a) and (b) (2006) Pharmaceutical cGMP regulations

Chapter 4
Statistical Concepts for the Analytical Chemist

Abstract Knowledge of statistics is very important for proper interpretation of the analytical data. The chapter on statistical concepts was included to identify and briefly review those concepts such as, significance, confidence intervals, uncertainty in measurements, design of experiments, etc. that are critical for the analytical chemist to understand so that they can discuss the data with company and regulatory agency statisticians and ensure that the data interpretation is acceptable for its intended use. It was not the intent for this discussion to be a complete statistical treatise, but the statistical concepts that will most likely be utilized by the analytical chemist in a GMP Analytical Chemistry Department. The necessary statistical equations are given with examples where deemed appropriate.

4.1 Distribution

When analytical data is collect and plotted as frequency against a variable, a distribution of the data can be observed. If the experiments are repeated many times and the data plotted, the distribution becomes more consistent and approaches a form which can be described by a mathematical equation. Most analytical data will approach a distribution function which is described as a "**Normal Distribution**". The normal distribution arises from the summation of many small random errors and can be expressed mathematically as $\int_{-\infty}^{\infty} \frac{1}{\sigma\sqrt{2\pi}} \left[e^{-\frac{(x-\mu)^2}{2\sigma^2}} \right] dx = 1$. It can be seen from the expression that the distribution is completely determined by the parameters μ the mean and the variance σ^2. The distribution is symmetrical about the mean, indicating where the highest probability of measurements occur, the probability of measurements occurring drops off sharply with 68.26 % occurring at $\mu \pm \sigma$, 95.44 % at $\mu \pm 2\sigma$, and 99.74 % at $\mu \pm 3\sigma$ [1].

The graph of the distribution is a bell-shaped curve (a.k.a. a normal curve) is shown in Fig. 4.1

T. Catalano, *Essential Elements for a GMP Analytical Chemistry Department*, DOI: 10.1007/978-1-4614-7642-9_4,

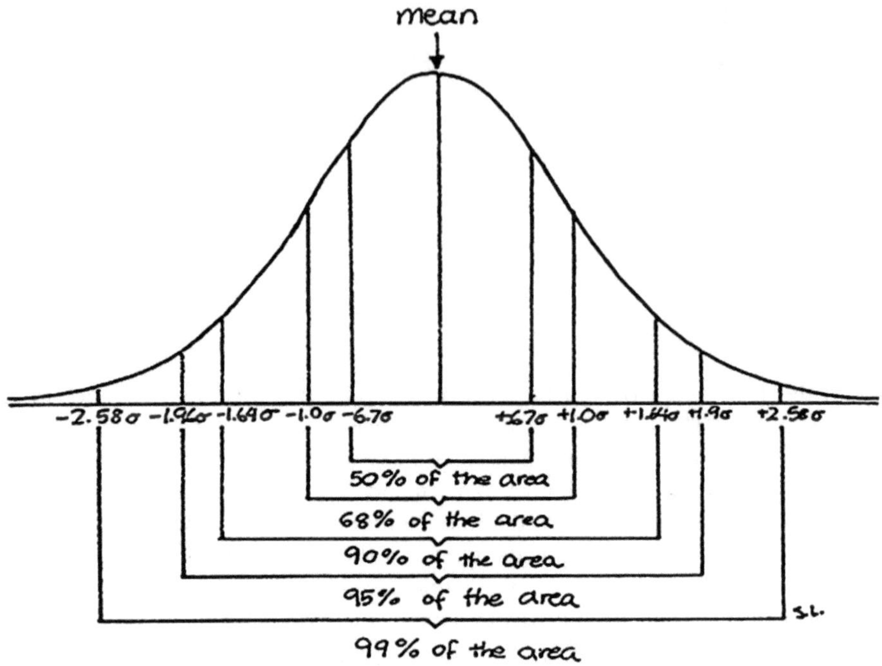

Fig. 4.1 Normal distribution

Distributions derived from the normal distribution are important for the analysis of analytical data, these distributions are described below.

4.1.1 Student's t Distribution

The t distribution is routinely used for determining if results have significant bias or for comparing observations with limits. The distribution is symmetrical and resembles a normal distribution and is described by the parameter "**Degrees of Freedom** (v)". As the number of degrees of freedom increase the distribution approaches the normal distribution. The t distribution is also used for determining confidence intervals which be discussed in a later section of the book.

The Degrees of Freedom refers to the number of independent pieces of data that has been used to measure a particular parameter. In general the degrees of freedom is based on the sample number (n) minus the number of parameters estimated from the data, for example the calculation of the standard deviation requires the determination of a parameter the mean (x), therefore the degrees of freedom utilized for the calculation of the standard deviation is $n - 1$ [1, 2].

4.1.2 Chi-Squared Distribution (x^2)

The Chi-squared distribution describes the distribution of estimates of variance. Specifically if the standard deviation of n data points is calculated the variance among the standard deviation has a Chi-squared distribution. The Chi-squared distribution is asymmetrical and become more symmetrical with an increase in v [1, 2].

4.1.3 The F Distribution

The F distribution describes the ratio of the variances. This is important when comparing the difference of variance of two data sets, it is also used in the analysis of variance (ANOVA). It is also very useful for comparing the precision of alternate methods of analysis. The F distribution is a ratio of two variances and is characterized by the degrees of freedom for each. The asymmetry of the distribution increases as the respective degrees of freedom decreases [1, 2].

4.2 Significance Testing

If two sets of data are determined to have different means X_a and X_b it is possible that both data sets come from the same population and that the difference observed was due to just random variation in the generation of the data. However, it is also possible that data generated came for different populations and their means are truly different and not due to just random variations. Significance testing can provide an approach which will allow for deciding which the more likely alternative is. Some basic statistic parameters are needed to be determined in order for significant testing to be performed [1, 2].

4.2.1 Arithmetic Mean

The mean X is the summation of all observations divided by the number of observations

$$\bar{x} = \frac{\sum_{i=1}^{n} X_i}{n} \tag{4.1}$$

4.2.2 Median

The Median is the central member of a series of observations arranged in ascending order. The median will have equal numbers of observations smaller and larger than its value. The median is generally a more robust value since it is less affected by extreme values, such as outliers. For example the series of observation 1, 5, 7, 8, 11, 12, 15, 20, 25 has a median value of 11.

4.2.3 Standard Deviation

The variance within a data set is the mean squared deviation of the individual values from the data set. The standard deviation is the positive square root of the variance. The variance and standard deviation are estimates of how individual values in the data set differ from each other, the larger the variance or standard deviation the larger the spread of individual data points within the data set.

$$\text{Variance } \left(S^2\right) = \frac{\sum_{i=1}^{n}(x_i - \bar{x})^2}{n - 1} \tag{4.2}$$

$$\text{Standard Deviation } (S) = \sqrt{\frac{\sum_{i=1}^{n}(x_i - \bar{x})^2}{n - 1}} \tag{4.3}$$

Although the standard deviation is a more commonly used parameter in analytical chemistry, it is important to know that when data is being considered for pooling the standard deviation of each estimate must be squared before summation.

4.2.4 Standard Error of the Mean

The standard error of the mean represents the variation of the mean. It represents the uncertainty which occurs from the random variation within an experiment. It is more precise than the standard deviation because it estimates the variation of averages.

$$\text{Standard Error of the Mean } S(\bar{x}) = \frac{S}{\sqrt{n}} \tag{4.4}$$

4.2.5 *Relative Standard Deviation (Coefficient of Variation)*

The relative standard deviation (RSD) or coefficient of variation (CV) is a comparison of the mean with spread in the data. The RSD is one of the most commonly used statistical parameter in analytical chemistry, it is usually expressed a percentage.

$$\%RSD = \%CV = \frac{S}{\bar{x}} \times 100 \tag{4.5}$$

4.2.6 *Hypotheses*

The null and the alternate hypotheses are the questions which must be asked when evaluating the significance of the data. The null hypothesis, which denoted as H_0, is interpreted as there is no difference between the data being compared and that the data is from the same population and the difference is due to random variation. The null hypothesis is expressed as follows:

$$H_0{:}u_A = u_B \tag{4.6}$$

where u = the population mean.

The alternate hypothesis, denoted by H_1, is interpreted as that the data being compared is different and is from different data populations and the difference in the data is not just due to random variation. This hypothesis is expressed as follows:

$$H_1{:}\ u_A \neq u_B \tag{4.7}$$

4.2.7 *Significant Testing [3]*

When performing significant testing the following steps should be considered.

1. State the null and alternate hypotheses.
2. Select the appropriate test statistic. In analytical chemistry the most common test statistics utilized are the *t* test and the F-test. The *t* test is used for the comparison of means and the F-test for the comparison of variances. In each test the calculated test statistic is compared to a critical value obtained from their distribution table at a particular level of probability for a determined level of degrees of freedom.
3. Choose whether the test statistic will be one-tailed or two-tailed. If the alternate hypothesis is H_1: $u_A \neq u_B$ than we are considering a two tailed statistic since we are concerned if there is a significant different for the mean of the data set in either directions. If we only want to consider whether the mean of one data set

Table 4.1 One sample t test calculation

Hypotheses	Test statistic calculation	Tails
H_0:u = u_0 H_1: u ≠ u_0	$t = \frac{\lvert\bar{x}-u_0\rvert}{S/\sqrt{n}}$	2
H_0:u = u_0 H_1: u > u_0	$t = \frac{\bar{x}-u_0}{S/\sqrt{n}}$	1
H_0:u = u_0 H_1: u < u_0	$t = \frac{u_0-\bar{x}}{S/\sqrt{n}}$	1

is significantly greater than the mean of the other data set the alternate hypothesis is H_1: $u_B > u_A$ or if the mean is significantly less than the mean of the other data set H_1: $u_B < u_A$ than we would consider a one-tailed statistic.

4. Choosing the level of significance is determined by what level of probability (α) is acceptable for difference between the data to be considered due to non-random error and be considered significantly different. In Analytical chemistry the most common level of probability chosen is $\alpha = 0.05$ level. This would suggest that the probability of values occurring at <0.05 would be considered to reject the null hypothesis and be considered a significant difference. If allowing 5 % of the values to reject the null hypothesis is too large than a lower value such as 0.01 should be used.
5. The t-statistic can be determine for the following comparisons:

   a. The comparison of the mean of a data set to a stated value (one sample *t* test)
   b. The comparison of means from two independent data sets (two sample *t* test).
      The one sample *t* test calculations are shown in Table 4.1.
      The two sample *t* test calculations are shown in Table 4.2.

However, the two sample t-statistic differs from the one sample t-statistic in that there are two data sets and therefore two standard deviations. Since there are two standard deviations one from each data set, the calculation requires the pooling of the standard deviations. If each data set has an equal number of observation the pooled standard deviations (S_{diff}) can be simplified to

$$S_{diff} = \sqrt{\frac{S_1^2 + S_2^2}{n}} \tag{4.8}$$

Table 4.2 Two sample t test calculation

Hypotheses	Test statistic calculation	Tails
H_0:u_1 = u_2 H_1: u_1 ≠ u_2	$t = \frac{\lvert X_1-X_2\rvert}{S_{diff}}$	2
H_0:u_1 = u_2 H_1: u_1 > u_2	$t = \frac{x_1-x_2}{S_{diff}}$	1
H_0:u_1 = u_2 H_1: u_1 < u_2	$t = \frac{X_2-X_1}{S_{diff}}$	1

Table 4.3 F test statistic calculation

Hypotheses	Test statistic calculation	Tails	DOF
$H_0{:}\sigma_a^2 = \sigma_b^2$ $H_0{:}\sigma_a^2 \neq \sigma_b^2$	$F = \frac{\sigma_{max}^2}{\sigma_{min}^2}$	2	ν_{max}, ν_{min}
$H_0{:}\sigma_a^2 = \sigma_b^2$ $H_0{:}\sigma_a^2 > \sigma_b^2$	$F = \frac{S_a^2}{S_b^2}$	1	ν_a, ν_b
$H_0{:}\sigma_a^2 = \sigma_b^2$ $H_0{:}\sigma_a^2 < \sigma_b^2$	$F = \frac{S_b^2}{S_a^2}$	1	ν_b, ν_a

If the data sets have unequal amounts of observation then the calculation of the two sample t-statistic becomes more complicated and statistician should be consulted.

6. Once the t-statistic is calculated it is compared to the critical value obtained from the t-distribution table at the chosen probability level, usually 0.05, with the degrees of freedom n – 1 for the one sample *t* test and the degrees of freedom of $n_1 + n_2 - 2$ for the two sample *t* test.

Another important statistic in analytical chemistry is the F-test statistic. The F-test is used for comparing variances s_a^2 and s_b^2 from two independent sets of data. Since variances are squared standard deviations it can also be used for comparing standard deviations, which is very useful for comparing the precision of analytical methods to see if one is significantly better than the other. The F-test statistic is calculated as a ratio of the variances s_a^2 and s_b^2 with larger value in the numerator and the smaller in the denominator. Since we are comparing variances from two independent data sets we need to utilize the degrees of freedom (ν_a, ν_b) for each of the variances. The F distribution table has the degrees of freedom listed on the top row for the numerator and the degrees of freedom for the denominator down the left side of the table. The calculation of the F-test statistic is shown in Table 4.3. The F = test statistic can be used to compare an observed variance to an expected or required variance [3]. In this circumstance the F-test calculation utilized is $F = \frac{S_b^2}{S_a^2}$ with the degrees of freedom $\nu_a = n - 1$ and the degrees of freedom $\nu_b = \infty$.

4.3 Confidence Intervals for the Mean [4]

The confidence interval is a range of values for the true mean u which would include the observed mean $\bar{x}$ with a given level of confidence. The 95 % confidence interval about the observed mean $\bar{x}$ is described by the following equation:

$$u - 1.96\left({}^{\sigma}/_{\sqrt{n}}\right) < \bar{x} < u + 1.96\left({}^{\sigma}/_{\sqrt{n}}\right) \tag{4.9}$$

The value 1.96 comes from the two tailed t value at $\alpha = 0.05$ and $v = \infty$. Typical analytical data consist of a relatively small number of data points which is used to calculate the mean $\bar{x}$. Therefore the population standard deviation σ and mean u is not known and must be replaced with the standard deviation S, and the mean $\bar{x}$. The 1.96 is replaced with two tailed t-statistic value with n − 1 degrees of freedom. The confidence interval for the mean equation is then revised to

$$\bar{x} \pm t\left({}^{S}\!/\!{}_{\sqrt{n}}\right) \tag{4.10}$$

4.4 Outliers in Analytical Data [4]

An outlier is an individual data point that is not consistent with rest of the data set. The outlier is usually observed as being distant from the remainder of the data set. These values have a large effect on calculated mean values and standard deviation values. Random variation can occasionally generate extreme values which are part of the valid data set and should be included in the data calculations. However, extreme data can also be produced by human error, analytical procedure aberration and instrument failure. These outliers should not be included in the data calculations so as to reduce their impact on the conclusions.

Outlier testing can be performed to identify outliers and distinguish them from data due to random variation and allowing for the suspected data to be investigated and corrected or eliminated from the data set. Visual inspection of the data will usually detect suspected outliers. The identification of an outlier by using outlier tests allow the analyst to direct their attention to problems and provide objective criteria for performing inspections or corrective actions. Outliers should not be removed from the data set solely on the bases of a statistical test; the decision should be based on the statistic and an investigation procedure such as a corrective action and preventive action (CAPA). It is important to realize that outliers are only relative to what is expected in the data set. It is important to consider that the outlier may be a relevant part of the data population. Consider a granular sample where some particles have much greater content of the analyte than the majority of other particles. In this case the outlier value is a relevant value and must be included in the data set. Statistical testing for outlier it is usually tested at both the 95 and 99 % confidence level. Outliers significant at the 99 % confidence level are usually rejected from the data set; however rejection of a large portion of the data is not permissible. Outliers significant at the 95 % confidence level are generally not rejected unless supported by other technical reasons.

Most analytical data sets are relatively small <50 observations; therefore the Dixon Q Test is commonly utilized as an outlier test. Applying the Dixon Q Test begins with ranking all of the data in ascending order x_1, x_2, $x_3 \ldots x_n$. Calculate the Q statistic for both the high and low outliers as described in Table 4.4:

When applying the Dixon Q statistic the recommend ranges of the data sets sizes should be followed, however going slightly beyond the recommend data set size is not usually serious. As the data set size increases there is a probability of

Table 4.4 Q statistic calculation

Test statistic	Application	Q table column (r_i) i = 10, 11, 12, 20, 21, 22
$\frac{x_2-x_1}{x_n-x_1}$ (low value) $\frac{x_n-x_{n-1}}{x_n-x_1}$ (high value)	Test for a single outlier value in a data set of n = 3–7	r_{10}
$\frac{x_2-x_1}{x_{n-1}-x_1}$ (low value) $\frac{x_n-x_{n-1}}{x_n-x_2}$ (high value)	Test for a single outlier value in a data set, unaffected by a single outlier value at the other end of the data set. Used for a data set of n = 8–10	r_{11}
$\frac{x_2-x_1}{x_{n-2}-x_1}$ (low value) $\frac{x_n-x_{n-1}}{x_n-x_3}$ (high value)	Test for a single outlier value in a data set, unaffected by up to two outlier values at the other end of the data set. Used for a data set of n = 5–10	r_{12}
$\frac{x_3-x_1}{x_n-x_1}$ (low value) $\frac{x_n-x_{n-2}}{x_n-x_1}$ (high value)	Test for a single outlier value in a data set, unaffected by one adjacent outlier value. Used for a data set of n = 5–10	r_{20}
$\frac{x_3-x_1}{x_{n-1}-x_1}$ (low value) $\frac{x_n-x_{n-2}}{x_n-x_2}$ (high value)	Test for a single outlier value in a data set, unaffected by one adjacent outlier value or an outlier value at the other end of the data set. Used for a data set of n = 11–13	r_{21}
$\frac{x_3-x_1}{x_{n-2}-x_1}$ (low value) $\frac{x_n-x_{n-2}}{x_n-x_3}$ (high value)	Test for a single outlier value in a data set, unaffected by one adjacent outlier value or up to two outlier values at the other end of the data set. Used for a data set of n = 14–30	r_{22}

two outlier values of masking each other. Following the recommend applications for each test will help avoid the masking of outliers and increase the probability of finding these aberrant values. Using several Q-test statistics on the same data set is not recommended.

4.5 Regression Analysis [5]

In analytical chemistry linear regression analysis is utilized to determine the relationship between two variables in a data set. The most common comparison is between the concentration of an analyte and its response from an analytical technique. For example, plotting the response from an analyte (dependent variable) on the y-axis and its concentration (independent variable) on the x-axis of a scatter plot will display the relationship between each variable. The purpose of regression analysis is to define the relationship in terms of a mathematical equation. If the relationship is believed to be linear the equation can be written as

$$y = a + bx \tag{4.11}$$

where b is the slope of the line and a is the intercept on the y-axis. The method of least-squares linear regression is use to determine the values of a and b for the best fitted line to the data. The best fitted line from the least-squares linear regression is determine by minimizing the sum of the squared differences between the observed values and the fitted values of y. The difference between the observed value and the fitted value ($\hat{y}$) is known as the residual. Before carrying out the least-squares linear regression calculation, a visual examination of the data should be performed. A scatter plot of the data should be prepared and examined for the appearance of an outlier or a disproportionate spread of the data. Ether of these conditions could have a significant effect on the position of the regression line, and affect the values of the slope and intercept of the regression line. The calculations of the slope b, and the intercept a, are as follows:

$$\mathrm{b} = \frac{\sum_{i=1}^{n}[(x_i - \bar{x})(y_i - \bar{y})]}{\sum_{i=1}^{n}(x_i - \bar{x})^2} \tag{4.12}$$

$$\mathrm{a} = \bar{y} - b\bar{x} \tag{4.13}$$

Other important statistics related to the least-squared linear regression is the residual, the residual standard deviation $(s_{y/x})$, the standard deviation of the slope (s_b) and the standard deviation of the intercept (s_a).

$$\text{Residual} = (\mathrm{y} - \hat{\mathrm{y}}) \tag{4.14}$$

$$s_{y/x} = \sqrt{\frac{\sum_{i=1}^{n}(y_i - \hat{y})^2}{n - 2}} \tag{4.15}$$

where

y_i The observed value.
$\hat{y}_i$ The calculated value of y from the regression equation.
n The number of pairs of data used the regression.

$$s_b = \frac{s_{y/x}}{\sqrt{\sum_{i=1}^{n}(x_i - \bar{x})^2}} \tag{4.16}$$

$$s_a = s_{y/x}\sqrt{\frac{\sum_{i=1}^{n} x_i^2}{n\sum_{i=1}^{n}(x_i - \bar{x})^2}} \tag{4.17}$$

Once the regression equation is applied and the values of the slope b, and intercept a, are determined it is important to understand the level of confidence of b, and a, so that accurate conclusions can be drawn from the data. The Confidence intervals about the data for the slope b, and the intercept a, are determine by Eqs. 4.18 and 4.19 respectively.

$$b \pm ts_b \tag{4.18}$$

$$\mathrm{a} \pm ts_a \tag{4.19}$$

where

t the two tailed t value at the desired significance level (usually 0.05) with degrees of freedom v = n – 2.

The measure of the linear relationship between the variables x, and y, can be determine by the correlation coefficient. The correlation coefficient (r) is calculated by utilizing Eq. 4.20.

$$\mathrm{r} = \frac{\sum_{i=1}^{n}[(x_i - \bar{x})(y_i - \bar{y})]}{\sqrt{\left[\sum_{i=1}^{n}(x_i - \bar{x})^2\right]\left[\sum_{i=1}^{n}(y_i - \bar{y})^2\right]}} \tag{4.20}$$

The value of r will be in the range of ± 1, when the value of $|r|$ is closest to 1 the greater the correlation between the variables exist. The correlation coefficient should not solely be taken as measure of linearity, it should utilized in conjunction with other information such as, the data of the independent variable being evenly distributed with no obvious aberrations and the plot of the residuals should appear to be randomly distributed with no apparent trends. If predictions are to be made from calibration curve produced from the regression equation the value of $|r|$ should be ≥ 0.9999.

Once the best fit straight line has been determined utilizing the regression equation, it is important to determine the uncertainty associated with the predicted value $\hat{x}$. The predicted value ($\hat{x}$) is calculated by the following equation:

$$\hat{x} = \frac{y_0 - a}{b} \tag{4.21}$$

where

y_0 The mean of N measurements of y.

The uncertainty for the predicted value $\hat{x}$ is calculated by Eq. 4.22.

$$s_{\hat{x}} \frac{s_{y/x}}{b} = \sqrt{\frac{1}{N} + \frac{1}{n} + \frac{(y_0 - \bar{y})^2}{b^2 \sum_{i=1}^{n}(x_i - \bar{x})^2}} \tag{4.22}$$

The value of $s_{\hat{x}}$ is referred to the as the standard error of the prediction for $\hat{x}$. The uncertainty of $\hat{x}$ is greatest at extreme ends of the data range and is at a minimum at the points $\bar{x}$, $\bar{y}$. The confidence interval for $\hat{x}$ is denoted by the equation:

$$\hat{x} \pm t s_{\hat{x}} \tag{4.23}$$

The determination of all of the above statistics for the least squared linear regression is important for interpreting the data. The confidence interval for the slope b, is not generally important when the regression line is being used as a calibration curve, since in the line should have a very high correlation for the variables x, and y, and the slope should be significantly different from zero.

However, when the regression line is used to determine limits, such as shelf life of a product than the confidence levels of the slope (b), become very important. In this case a wide confidence limit of the slope may cause a short dating of the products shelf life.

It is important to determine the linearity of the calibration line. As previously mentioned the correlation coefficient is not a very good for the determination of linearity, more specific tests for non-linearity can be utilized. Non-linearity can be detected by observing the data on a scatter plot, specifically a plot of the residuals. Comparing the residual standard deviation $\left(s_{y/x}\right)$ with the standard deviation of y $\left(s_y\right)$ values from multiple observations for single value of x, and utilizing the F statistic can be a good indicator of linearity or the presence of non-linearity. The F statistic is calculated from the following equation:

$$F = \frac{{s_{y/x}}^2}{{s_y}^2} \tag{4.24}$$

The null hypothesis is H_0: $s_{y/x} = s_y$ and the alternate hypothesis, H_1: $s_{y/x} > s_y$. The F test is therefore a one tailed statistic with n − 2 degrees of freedom for $s_{y/x}$, where n is the number of data pairs in the regression data set and n − 1 degrees of freedom for s_y where n is the number of replicate observations for the single value of x. The critical value for F is obtained from the F distribution table for a chosen probability α (usually 0.05), with v_1 degrees of freedom for $s_{y/x}$ and v_2 degrees of freedom for s_y. If the calculated value of F is greater than the critical value found in the table, than the null hypothesis is rejected and the residuals are significantly different than can be attributed to random variation. This result along with a plot of the residuals demonstrating a non-random trend would be indicative of non-linearity. Another approach to evaluating linearity is to determine the fit of the data to a polynomial regression. If the data has a better fit to polynomial regression it is an indication that there is curvature to the data and there is significant departure form linearity. The determination of polynomial regression will not be discussed in this book but can be found in many text books on statistical regression analysis.

Once the calibration line has been determined to have a satisfactory linear fit to the data, it is important to determine if the intercept (a), is significantly different form zero. This can be determined by calculating the confidence interval of the intercept and determine if zero in included in the interval. If it is found that zero is included within the confidence interval than the intercept is not significantly different from zero (which is the desired result). If it is found that zero is not included in the confidence interval of the intercept than the intercept is significantly different from zero, which would indicate some bias is influencing the regression line and an investigation of the data should be performed.

4.6 Design of Experiments [6–8]

Statistical design of experiments has become very popular in analytical chemistry. It is being utilized for method development, method validation, robustness, stability study design, quality by design, etc. The advantage of statistical design of experiments is that you can evaluate the effect of several variables individually or in combination on the process at the same time. Before we could discuss statistical design of experiments we need to understand the terminology used. The following terms are of particular importance.

Factor	A variable or cause which can affect the results of a process.
Treatment	The run of an experiment utilizing particular factors.
Response	The measurement of a particular output from the process.
Effect	The change in the response caused by the change in the level of factors.

An important result from the utilization of design of experiments is the cost saving by obtaining the same data while performing much fewer measurements. Factorial designs are among the most popular in analytical chemistry. A design where two or more factors are investigated and all combinations of levels appear in the experiment is known as a full factorial design. Factorial designs that have the same number of levels for each factor is abbreviated as, f factors each with n levels, n^f factorial. For example a two level design with three factors is designated as 2^3 full factorial design. Factorial designs are most commonly used to determine significance of changes due to experimental parameters or factors. The number of experiments to be performed grows very rapidly as the number of factors or levels increase. For example at two level two factor design requires $2^2 = 4$ number of treatments while a four factor design requires $2^4 = 16$ treatments. Generally, additional sample replication is required which adds to the number of treatments. For full factorial designs greater than four factors, a fractional factorial design approach is utilized. Fractional factorial designs allow for the evaluation of several factors with a lower number of treatments, than a full factorial design. The reduction in treatments does come with a price, the loss of some information. However the information lost, usually higher order interaction factors, which are usually less important is a good trade-off for the reduced number of treatments. The information loss is due to the concept known as confounding or aliases. When a fractional factorial design is implemented some of the main factors will be confounded, meaning that the effects determined are a combination with other effects (usually interaction terms). Since the interaction terms are usually much less than the main factor effect, value of the main factor effect is not significantly impacted by the interaction effect. The fractional factorial design is designated by the n^{f-k}, where n is number of levels, f the number factors for a full factorial design, and k the number of factors remove from the design. For example a 2^3, full factorial design containing eight treatments could be done with four

Table 4.5 2^3 Full factorial design

Run	A	B	C	AB	AC	BC	ABC
1	+	+	+	+	+	+	+
2	–	+	+	–	–	+	–
3	+	–	+	–	+	–	–
4	–	–	+	+	–	–	+
5	+	+	–	+	–	–	–
6	–	+	–	–	+	–	+
7	+	–	–	–	–	+	+
8	–	–	–	+	+	+	–

treatments by utilizing a fractional factorial design 2^{3-1}. An example for a 2^3 full factorial design is described in Table 4.5. The "+" represents the high level and the "–" represents the low level of the factors. Factors AB, AC, BC and ABC are second and third order interaction factors respectively. Table 4.6 describes a 2^{3-1} fractional factorial design, where the runs with negative values for the ABC interaction are removed from the full factorial design. It can be seen that some information loss occurred from the reduction of the design from 8 runs to 4 runs. The consequence of doing the fractional factorial design results in the confounding of factors A, B and C with interaction terms BC, AC, and AB respectively, however in most cases the effects of these interaction terms are very minor compared to the effects of the main factors so it was well worth performing only half the number of runs.

There are many more advanced designs which are appropriate for specific applications, such as Blocked designs, Mixture Designs, Latin Square, Star, Central Composite, to mention a few. The implementation of the designs requires software programs which are addressed in many textbooks and web sites. Several references are available in the reference section of this chapter.

4.7 Required Sample Replicates [4]

One of the most common asked questions by analytical chemist is, how many replicate samples do I need for this experiment? The answer is not a simple response, it depends on a desired limit, the level of confidence required (usually 95 %) and the acceptable uncertainty level of the process (usually a 3:1 ratio of

Table 4.6 2^{3-1} Fractional factorial design

Run	A	B	C	AB	AC	BC
1	+	+	+	+	+	+
2	–	–	+	+	–	–
3	–	+	–	–	+	–
4	+	–	–	–	–	+

process to assay standard deviation) which is represented by the standard error of the mean $S(\bar{x}) = \frac{S}{\sqrt{n}}$ Eq. 4.4. The minimum number of replicates can be calculated by the following equation.

$$n_{min} \geq \left[\frac{s}{s(\bar{x})}\right]^2 \tag{4.25}$$

Example: If a formulation process has a standard deviation of 4 % and desires to produce their lots within a specification of ±5 % at 95 % confidence level, what is the minimum number of replicates samples required?

Confidence interval = $t(S(\bar{x}))$.

t = table value for n − 1 degrees of freedom for the 95 % confidence level.

If n = 6 than $S(\bar{x}) = \frac{5}{2.776} = 1.80$

$$n_{min} \geq \left[\frac{4}{1.80}\right]^2 = 4.9.$$

This would be raised to five replicates.

In some circumstances you have a required number of replicates you can analyze, so the information you may need to know what the coefficient of variation (CV) required to meet a specification at the 95 % confidence interval. To calculate the required CV the following equation can be utilized.

$$\%CV = \frac{\%expected\ difference \times \sqrt{n}}{t_{value}} \tag{4.26}$$

where

Expected difference	Specification value.
t_{value}	Two tailed value in the t distribution table for n − 1 degrees of freedom.
n	the number of replicate samples.

Example: If a formulation process has a desire to produce their lots within a specification of ±5 % at a 95 % confidence level, and they are planning to use two replicate sample analysis. What is the %CV required by the analysis method.

$$\%CV = \frac{5 \times \sqrt{2}}{12.706} = 0.57\,\%$$

Table 4.7 illustrates the %CV required for several of the most common specifications at the 95 % confidence level.

Table 4.7 Percentage of CV required for specifications with various sample replicates at 95 % C.I

Specification Replicates	90–110 %	95–105 %	98–102 %
2	1.1	0.57	0.24
3	3.9	2.0	0.8
6	6.5	4.6	1.9
10	13.7	6.9	2.8

4.8 Method Performance

Method validation is the process which is utilized to determine if a method performance is appropriate for its intended use. Method validation consists of evaluating parameters such as, accuracy, precision, linearity, limit of quantitation, limit of detection, specificity, and ruggedness. The extent of method validation performed is based on the stages of development as shown in Sect. 3.2.3.

4.8.1 Method Precision [4]

Method precision is defined as closeness among individual measurement for a specified sample. Precision is usually determined by the calculation of the standard deviation (Eq. 4.3) or relative standard deviation (Eq. 4.5) obtained from replicate measurements of an appropriate sample. Method precision is evaluated as two entities, repeatability and intermediate precision. The conditions under which the measurements are made determine which type of precision is being estimated.

Repeatability or also referred to within run precision is performed by a single analyst, on a single instrument, during a single run. This type of precision is an estimate of the variation among replicate measurements in a single run using the same sample throughout the run.

Intermediate precision is the determination of precision of replicate measurements utilizing more variable conditions then repeatability. In the estimation of intermediate precision variation such as different runs or days, different analysts, different sets of equipment, or any other variables that may occur during the routine implementation of the method. It is essential that all the variations applied during the determination of intermediate precision be documented. The variances for repeatability and intermediate precision are shown in Table 4.8.

Utilizing the variances from Table 4.8. Repeatability and Intermediate Precision is calculated as follows:

$$\text{Repeatability } (\mathrm{P_r}) = \sqrt{MS_w} \tag{4.27}$$

Table 4.8 Variance for k groups with n replicates per group

Variance	Sum of squares	Degrees of freedom	Mean square	F statistic
Within group (S_w)	$S_w = \sum_{i=1}^{k} (x - \bar{X}_i)^2$	N − k	$MS_w = \frac{S_w}{N-k}$	$\frac{MS_b}{MS_w}$
Between group (S_b)	$S_b = \sum_{i=1}^{k} (\bar{x}_l - \bar{x}_{GM})^2$	k − 1	$MS_b = \frac{S_b}{k-1}$	
Total (S_{tot})	$S_{tot} = S_w + S_b$	N − 1		

K Number of treatment groups
N Number of samples per group
N Total number of samples
$\bar{X}_{GM}$ Grand Mean for total number of samples (N)
S_w Sum of squares within group
S_b Sum of squares between groups
S_{tot} Total sum of squares
MS_w Mean square within
MS_b Mean square between

An estimation of the Between Group precision is as follows

$$\text{Between } (P_b) = \sqrt{\frac{MS_b - MS_w}{n}} \tag{4.28}$$

An estimation of the Intermediate Precision is as follows

$$\text{Intermediate Precision } (P_I) = \sqrt{P_r^2 + P_b^2} \tag{4.29}$$

Reproducibility is determined using the same calculations as intermediated precision, except that the variations are between two different laboratories. In many cases reproducibility can be substituted for intermediate precision (see ICH 2Q(R)).

If the precision of the methods need to be compared to determine if the methods precision is significantly different from each other, the F statistic can be calculated using the equations in Table 4.3.

4.8.2 Accuracy

Accuracy is defined as the closeness of a measurement to the true value. Therefore the measurement includes both precision and bias as part of its value. Trueness is usually express in terms of bias. Bias is can be evaluated when the mean ($\bar{x}$) of several measurents is compared to the true value (u_0). In practice the true value is a certified reference value or reference standard spiked in the sample matrix. Bias is therefore calculated as

$$\text{Bias } = \bar{x} - u_0 \tag{4.30}$$

$$\%\text{Bias} = \frac{\bar{x} - u_0}{u_0} \times 100 \tag{4.31}$$

$$\text{Recovery } (\%) = \frac{\bar{x}}{u_0} \times 100 \tag{4.32}$$

It is usually important to determine if the mean value ($\bar{x}$) is significantly different from true value (u_0). This can be calculated using the student t test at a specified confidence level, usually 95 %, as described in Table 4.1. If there is a significant difference between obtained mean value ($\bar{x}$) and the true value (u_0) the measurement is significantly biased should not be used for accuracy determination.

4.8.3 Linearity [1, 4]

The evaluation of linearity is essential for determining if the instrument response is linear with the analyte concentration. The linearity is determined by the evaluation of several concentration levels, generally not less than five, equally spaced over a concentration range utilizing least square linear regression analysis and the correlation coefficient as discussed in Sect. 4.5. Once linearity is established a working concentration range must be determined to establish that the method is adequate for its intended use. The working range must include the LOQ and at a minimum 10 % above the target concentration. The samples used to evaluate the working concentration range must mimic the sample matrix of material which will be analyzed by the method, such as certified reference standard, spiked placebo, or prepared matrix matched standard solutions. If non- linearity is observed with the matrix sample studies, this may indicate the presence of interfering compounds or other bias present in the method, in any case the method needs to be investigated.

4.8.4 Limit of Detection (LOD) [4]

The LOD is the minimum concentration of an analyte that can be detected that is significantly different than the response of the blank. The LOD can be estimated by obtaining the standard deviation of replicate analysis, of a blank sample (no analyte present in the sample). Generally 6–10 replicates taken through the analytical method should be obtained. Using statistical reasoning a limit which only allows 5 % of the distribution values to be considered a false positive result can be calculated by multiplying the standard deviation by a factor based on a one tailed student t value, with infinite degrees of freedom, and adding that to the mean value x_0, of a blank. Thus choosing α at 0.05 the t value is 1.65 and the limit value is $x_0 + 1.65\sigma_0$. Therefore any value equal to or greater than the limit value is considered a positive finding and the value is statistically above zero. However, we also have to consider values that appear to be negative, but may in fact be positive (false negative). Using the same statistical reasoning a concentration can be determined which the limit value will cut off an area β of the expected distribution for the calculated concentration. This new, higher, calculated concentration is the called the **Limit of Detection**. Generally the significance level for α and β is the same, 0.05, allowing for a 5 % false negative rate. The limit of detection for a method where no correction for baseline is performed can be shown in the following equation:

$$\text{LOD} = x_0 + \sigma t_{\alpha,v} + \sigma t_{\beta,v} \tag{4.33}$$

$$\text{LOD} = x_0 + 1.65\sigma + 1.65\sigma = x_0 + 3.3\sigma$$

If baseline corrected, $x_0 = 0$

$$\text{then LOD} = 3.3\sigma$$

where

x_0 Mean of blank measurements.

σ Standard deviation of the blank.

T the value of the one tailed student t with infinite degrees of freedom.

To compare the calculated LOD from detectors with different responses, the LOD can be normalized by dividing the LOD with the slope of the calibration line. The more general equation for LOD than is as follows:

$$\text{LOD} = \frac{3.3\sigma}{S} \tag{4.34}$$

where

S Slope of the analyte calibration line.

The Eq. 4.34 for LOD is consistent with the equation for LOD within the ICHQ2(R) guideline.

4.8.5 *Limit of Quantitation (LOQ)*

The LOQ is the lowest level of an analyte that can be quantitated with a specified of confidence. Similar to approach for the LOD, the standard deviation of multiple replicates of the blank, σ, is utilized in the calculation. A value of 10σ is frequently used. Also to normalize LOQ among detectors of varying responses the LOQ is divided by the slope of the analyte calibration line, thus resulting in a general equation for LOQ as follows:

$$\text{LOQ} = \frac{10\sigma}{S} \tag{4.35}$$

4.8.6 Ruggedness [8, 9]

Ruggedness of a method is preformed to evaluate the effect of small changes in the methods parameters such as, temperature, pH, flow rate, solvent composition, sample size, etc., on the measurement result. The goal is to identify variations in method conditions that can significantly impact the measurement results due to analysis being performed at different times or in different laboratories. Ruggedness can be evaluated by testing each parameter independently however this can be labor intensive if there are many parameters. Also, using a univariant approach will not give any information about interactions among the parameters. Ruggedness testing usually employees the use of fractional factorial designs as described in Sect. 4.6. A particular design known as the **Plackett-Burman** is generally chosen since it can measure seven independent parameters (factors) with eight experiments. Factors identified to have a significant effect measurement result, will need to further investigated. One solution for controlling factors identified to have a significant effect on the measurements is to tighten their control limits. If this is not successful than further method development will be required. Another important outcome from ruggedness testing is the identification of critical parameters which should be included in the system suitability criteria. The Plackett-Burman fractional factorial design is shown in Table 4.9.

The Plackett-Burman design is a balanced design so that the results of each experiment are split into two groups of four for each parameter, one group for high level (+) and one group for low level (−). To determine the effect of a parameter, calculate the difference between the average of the results for the (+) level and the average of the results for the (−) level. For example the effect of parameter A (E_A) is calculated as follows:

$$(E_A) = \frac{s+t+u+w}{4} - \frac{v+x+y+z}{4} \tag{4.36}$$

We now need to determine which parameter effects ($E_{A\text{–}G}$) will have a significant impact on the Method performance. The absolute value of the effects

Table 4.9 Plackett-Burman design

Run	A	B	C	D	E	F	G	Result
1	+	−	−	+	−	+	+	s
2	+	+	−	−	+	−	+	t
3	+	+	+	−	−	+	−	u
4	−	+	+	+	−	−	+	v
5	+	−	+	+	+	−	−	w
6	−	+	−	+	+	+	−	x
7	−	−	+	−	+	+	+	y
8	−	−	−	−	−	−	−	z

$|E_{A-G}|$ can be evaluated statistically by calculating a critical value, E_{crit}. The effect critical value is determined as shown in Eq. 4.37

$$E_{crit} = \frac{ts}{\sqrt{2}} \tag{4.37}$$

where

s The standard deviation (Precision) of the method.

t The two tailed student t value for the desired level of confidence (usually 95 %) with the degrees of freedom equivalent to that used in the standard deviation (s) determination.

Any values of parameter effects $|E_{A-G}|$ which are greater than E_{crit} are considered to have a significant impact on the method performance.

4.9 Measurement Uncertainty [4, 10]

Measurement is a process, in which specified procedures are performed to determine a value. In a measurement process even when all the measurement factors are controlled, repeated observation using the same process under the same condition, are rarely found to be identical. This is due to the variables such as, operator, reference standards, materials, instrument, environment, calibration, and test methods. Therefore measurement results are never the true value because of the uncertainty associated with them. The following list contains the need for the determination of uncertainty.

a. The customer needs to know that measurement uncertainty has to be taken into account particularly when regarding specification limits.
b. Testing laboratories should include the uncertainties with their calibration certificates.
c. Testing laboratories shall have procedures for estimating uncertainty of measurement.
d. The uncertainty values that may be required in the report of analysis depends on requirements by the test method, or whether conformance to specification has to be assessed.
e. In calibration, uncertainties have to be stated in the certificate of analysis so that it can be utilized by the user.

There are several essentials of measurement uncertainties that must be considered. See Fig. 4.2.

There are two methods of estimating uncertainty. Type A, where the estimation is based on statistical analysis of replicate measurements. Type B, estimations come from any other sources such as uncertainty value in reference standard, value in a calibration certificate, specification on a volumetric flask or balance. Type A

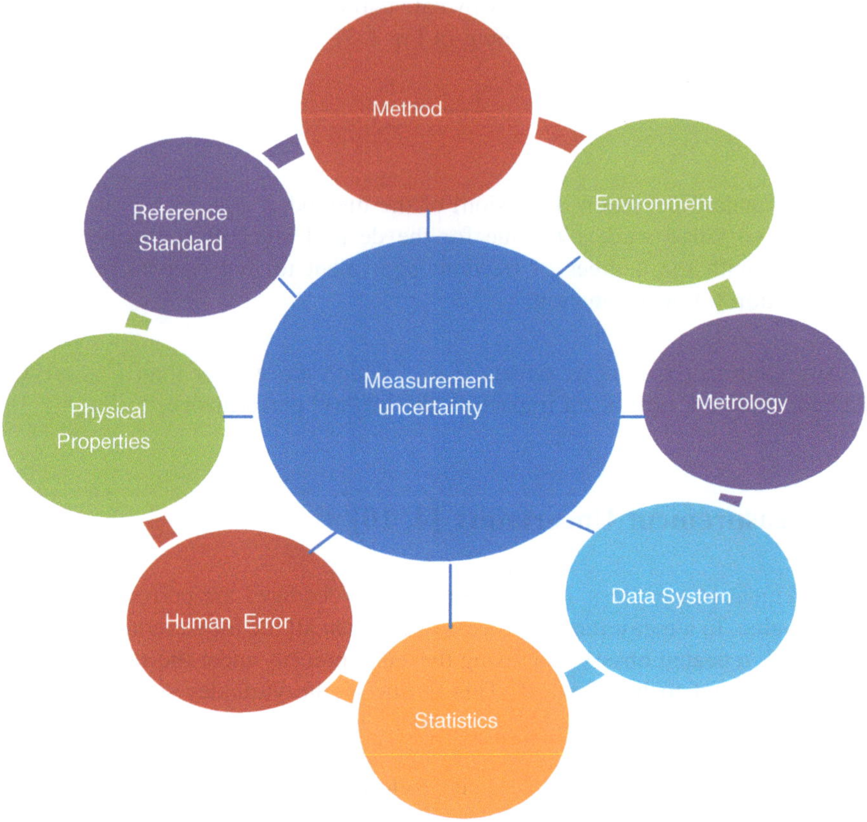

Fig. 4.2 Contributions to the measurement of uncertainty

and Type B estimations are used in the same way when combining their uncertainties. Individual uncertainty is known as a **standard uncertainty (u)** and a reported uncertainty is known as an **expanded uncertainty (U)** [11].

There are two rules for calculating uncertainties in the analytical result.

1. If the quantities are added or subtracted to obtain a result y, the uncertainty in y is the uncertainty $u(x_i)$ in x_i.
2. If the quantities are multiplies or divides, the expression for y the contribution of uncertainty in y, u(y)/y is the uncertainty $u(x_i)/x_i$ in x_i.

In more general cases, where the result is due to several different Algebraic operations the uncertainty must be obtained by a different approach. In this case the approach is to determine the change in the result, y based on the uncertainty in, x. That is the uncertainty change in x_i times the rate of change in, y with x_i. This results in the uncertainty calculation for $u_i(y)$ in y, with an uncertainty $u(x_i)$ in x_i. This is shown in Eq. 4.38.

$$u_i(y) = c_i u(x_i) \tag{4.38}$$

where

c_i The slope of the line of y against x_i.

Utilizing Eq. 4.38 for the combination of uncertainties, when the contributions are independent of each other, they combine as the root sum of their squares. This is shown in Eq. 4.39.

$$u(y) = \sqrt{\sum_{i=1}^{n} c_i^2 u(x_i)^2} \tag{4.39}$$

Table 4.10 describes the combination of uncertainties for independent contributions.

Combining uncertainties from contributions which are not independent from each other requires a much more complex relationship, which can be found in [6]. Reporting the measurement uncertainty is usually as an expanded uncertainty, U. The expanded uncertainty, U is determined by multiplying the standard uncertainty, u by a factor k. The determination of factor k is based on the student t value at a desired confidence level (usually 95 %). For a confidence of 95 % (0.05) the t value at infinite degrees of freedom is 1.96 and the k factor used is rounded to 2. Therefore the reported uncertainty U will be two times the standard uncertainty u. The resulting value, U is commonly rounded up to one or two significant figures.

4.10 Sampling Strategies [2, 4]

Sampling is one of the most important processes to insure that the data obtained from the analysis is appropriate for its intended use. Improper sampling can contribute the greatest amount of uncertainty to the measurements. Therefore planning of a sampling protocol which is suitable for the intended use of the method must be developed and executed. From a statistical aspect sampling is a subset of items taken from a larger population with minimum bias and is representative of that population. There are many different types of samples, some of which will be discussed in this section.

Table 4.10 Combination of uncertainties

Calculation of result, y	Uncertainty u_y in y from x_i
$y = x_1 + x_2$ $y = x_1 - x_2$	$u_y = \sqrt{u_{x_1}^2 + u_{x_2}^2}$ $u_y = \sqrt{u_{x_1}^2 - u_{x_2}^2}$
$y = x_1 \times x_2$ $y = x_1/x_2$	$u_y y = \sqrt{\left(u_{x_1} x_1\right)^2 + (u_{x_2} x_2)^2}$ $\frac{u_y}{y} = \sqrt{\left(\frac{u_{x1}}{x_1}\right)^2 + \left(\frac{u_{x2}}{x_2}\right)^2}$

Bulk Sample—The primary material that requires characterization or analysis such as, the lot or batch of a drug substance or a drug product.

Sub-Sample—A selected portion of a larger sample. This can be the removal of selective samples from the bulk, by the use of a device known as a thief or the removal of samples from a laboratory sample for analysis.

Composite Sample—The combining of a collection of sub-samples into a uniform homogenized sample for analysis. When applicable it serves as a reduction in use of analytical resources.

There are several strategies to sampling that must be considered so that the sample analysis is adequate for its intended use. It is important that each sample taken has an equal probability of containing all the properties of the bulk. Randomization of sampling is a strategy which is usually used as the best insurance against obtaining unwanted biases. Although randomization can minimize the bias among the sub-samples it does not ensure that the samples are representative of the bulk material. Therefore it is necessary to obtained samples that are unbiased but also have a sufficiently small variance for their intended use. There are several sampling strategies which have different variances. It is the goal to select those strategies which deliver small variances along with randomization procedures which provide unbiased results.

4.10.1 Simple Random Sampling

Simple random sampling provides the probability that every sub-sample chosen has an equal chance of containing all of the properties found in the Bulk sample. In the case of choosing items from a number of discrete items such as, drums, packages, bottles etc., random number generators are used to choose the items in a random order. For particulate matter such as drug substance and solid drug product the technique of repeated quartering is utilized, but care has to be taken not to increase the bias due to segregation of particle size or other physical properties which are observed. Simple random sampling is among the easiest to implement, however because of the variation in the intervals between items chosen it would be a poor choice for sampling from a continuous process. The reason being if the material is non-homogeneous or has physical properties which are changing, the variance from simple random sampling would be among highest of the sampling strategies.

Statistical treatment for simple random sampling is dependent on the number of sample, n taken from the sample population, N. The standard error of the mean $s(\bar{x})$ due to sampling is given by Eq. 4.40

$$s(\bar{x}) = s_{sam}\sqrt{\frac{1-f}{n}} \tag{4.40}$$

where

s_{sam} The standard deviation of the sampling process.

f n/N.

The standard deviation of the sampling process, s_{sam} can be obtained from the between group component of variance described in Table 4.8. When N is very large or the sampling n, is less than 10 % of the population N, than f can be ignored and Eq. 4.40 is reduced to Eq. 4.41.

$$s(\bar{x}) = \frac{s_{sam}}{\sqrt{n}} \tag{4.41}$$

When dealing with particulate matter the sampling statistics can also be treated as a large N situation and Eq. 4.41 can be utilized.

If it is determined that analytical uncertainty is very small than s_{sam} is the standard deviation (s) of the observation x_i.

4.10.2 Stratified Random Sampling

In the stratified random sampling process the sampled population is divided into segments and each segment is sampled as per the simple random sampling process stated in Sect. 4.10.1. The number of items selected from each segment is dependent on the intended use of the measurement. There are several considerations that must be addressed when applying the stratified random sampling process.

- There must be equal number of items per segment and the same number of items is chosen from each segment. Having each segment of equal size, allows the use of composite sampling of the whole without introducing a biased estimate of the bulk composition.
- Proportional sampling is the process of taking the number of items from each segment proportional to the fraction of each segment in the bulk. That is if the bulk has samples that are different, than random samples from each segment would be chosen such that the total samples taken are in proportion with each of the types of samples in the bulk. Proportional sampling also provides the opportunity for composite sampling. The variance observed from proportional sampling is smaller than that produced from simple random sampling, especially when there is no information available about the variances of the individual segments.
- Optimal allocation is utilized to determine the number of samples per segment, based on the size and the standard deviation of each segment. This strategy provides the smallest variance for the total number of samples in the segment. The optimum number samples (n_i^{opt}) to take from the ith segment of n samples are determined by Eq. 4.42.

$$n_i^{opt} = \frac{P_i\sigma_i}{\sum P_i\sigma_i}\text{n} \tag{4.42}$$

where

P_i The proportion of segment i in the population.

σ_i The standard deviation of samples in the ith segment.

The statistical treatment for stratified random sampling requires the determination of the mean $\bar{x}_i$, total mass in each segment m_i, the standard deviation $s_{sam,i}$ and the number samples in the segments n_i. The proportion, P_i of each segment in the bulk is shown below.

$$P_i = \frac{m_i}{\sum_{i=1}^{n_i} m_i} \tag{4.43}$$

$$\text{The mean of the Bulk } \bar{X} = \sum_{i=1}^{n_i} P_i s_{sam,i} \tag{4.44}$$

$$\text{The variance } s^2(\bar{X}) = \sum_{i=1}^{n_i} \frac{P_i^2 s_i^2 (1 - f_i)}{n_i} \tag{4.45}$$

$$\text{The standard error of the mean } s(\bar{X}) = \sqrt{s^2(\bar{X})} \tag{4.46}$$

$$\text{The total mass of analyte} = m_{tot}\bar{X} \tag{4.47}$$

$$\text{Standard uncertainty from sampling} = m_{tot} s(\bar{X}) \tag{4.48}$$

References

1. Freund JE (1967) Modern elementary statistics, 3rd edn. Prentice Hall, Englewood Cliff
2. Zar JH (1984) Biostatistical analysis, 2nd edn. Prentice Hall, Englewood Cliff
3. Mandel J (1964) The statistical analysis of experimental data. Dover Publications Inc, New York
4. Ellison SLR et al (2009) Practical statistics for the analytical scientist, 2nd edn. RSC, Cambridge
5. Draper N, Smith H (1981) Applied regression analysis, 2nd edn. Wiley, New York
6. Box GEP, Hunter WG, Hunter SJ (1978) Statistics for experimenters: an introduction to design, data analysis, and model building. Wiley, New York
7. Barrentine LB (1998) An introduction to design of experiments: a simplified approach. ASQ Quality Press, Milwaukee
8. Deminig SN, Morgan SL (1987) Experimental design: a chemometric approach. Elsevier, Amsterdam
9. Catalano T (1995) Chemometric techniques in analytical chemistry, Searle, Worldwide analytical workshop
10. ISO Guide 98 (1995) Guide to the expression of uncertainty in measurements. International Organization for Standardization, Geneva
11. ISO 3534-2 (2006) Vocabulary and symbols, Part 2: Applied statistics. International Organization for Standardization, Geneva

Chapter 5
Outsourcing

Abstract There are many reasons why companies outsource development activities such as, not having the capabilities in-house, their expertise is limited to a specific area, or may have certain resource capability gaps that require the need for outsourcing. Others may be seeking novel technology or unique capabilities offered by an outsource provider. Tight timelines, competing priorities, or the need to allocate resources for other strategic purposes may also lead to a decision to outsource. In many occasions outsourcing is performed on an individual basis which leads to many concerns such as, higher costs, poor responsiveness, lack of partnership relations etc. The development of an outsourcing process would provide the most efficient and effective approach to outsourcing. In this chapter an outsourcing process is described which can be applied to small and large companies alike. The flow charts describing the process and the documentation needed are included so that a clear and complete understanding can be obtained.

5.1 Process

The pharmaceutical development process is a complex and dynamic one, requiring detailed knowledge of Drug Substance characterization, Product Development, Analytical Support, and regulatory requirements. Given the expertise necessary to successfully develop a pharmaceutical product, many companies opt to outsource development activities. There are many reasons why companies outsource development activities such as, not having the capabilities in-house, their expertise is limited to a specific area, or may have certain resource capability gaps that require the need for outsourcing [1]. Others may be seeking novel technology or unique capabilities offered by an outsource provider. Tight timelines, competing priorities, or the need to allocate resources for other strategic purposes may also lead to a decision to outsource. The recent increases in pharmaceutical companies

T. Catalano, *Essential Elements for a GMP Analytical Chemistry Department*,
DOI: 10.1007/978-1-4614-7642-9_5,

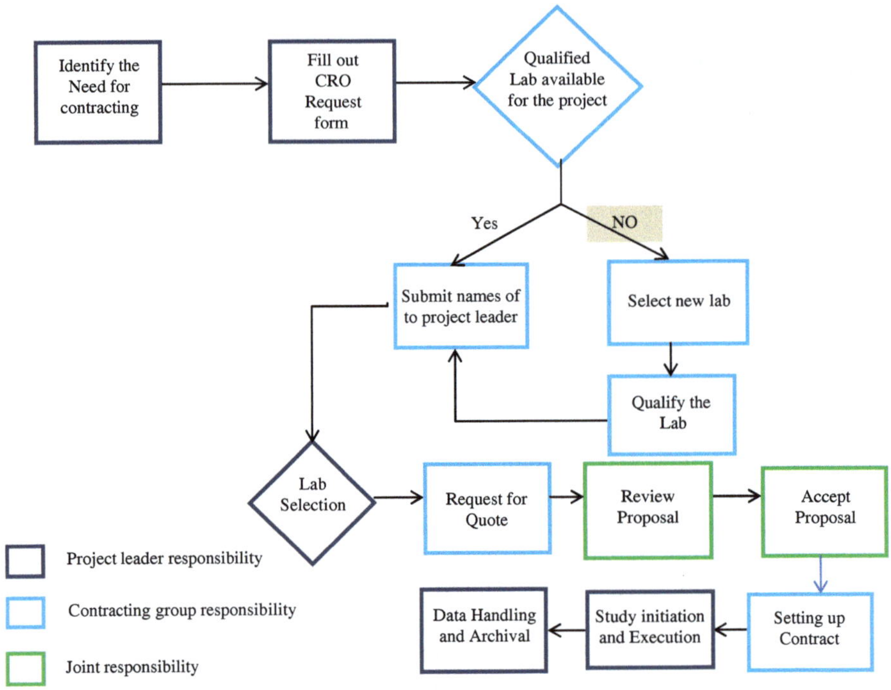

Fig. 5.1 Outsource process flow diagram

mergers and acquisitions, may drive the need to outsource in order to keep existing projects on track while integration activities are being implemented. In addition many companies would like to minimize the use of their scientific resources for routine or non-scientific activities. For example the sponsor company may design a study, such as stability, but outsource the implementation of the testing which would be considered routine.

The implementation of outsourcing activities within a company should be performed in an efficient and effective manner. In many occasions outsourcing is performed on an individual basis which leads to many concerns such as, higher costs, poor responsiveness, lack of partnership relations etc. The development of an outsourcing process would provide the most efficient and effective approach to outsourcing. Depending on the size of the company or the extent of outsourcing required, the process that is developed must be owned by an individual, known as the process owner, or contracting group responsible for managing the outsourcing. Figure 5.1 describes a basic overview for an outsourcing process [2].

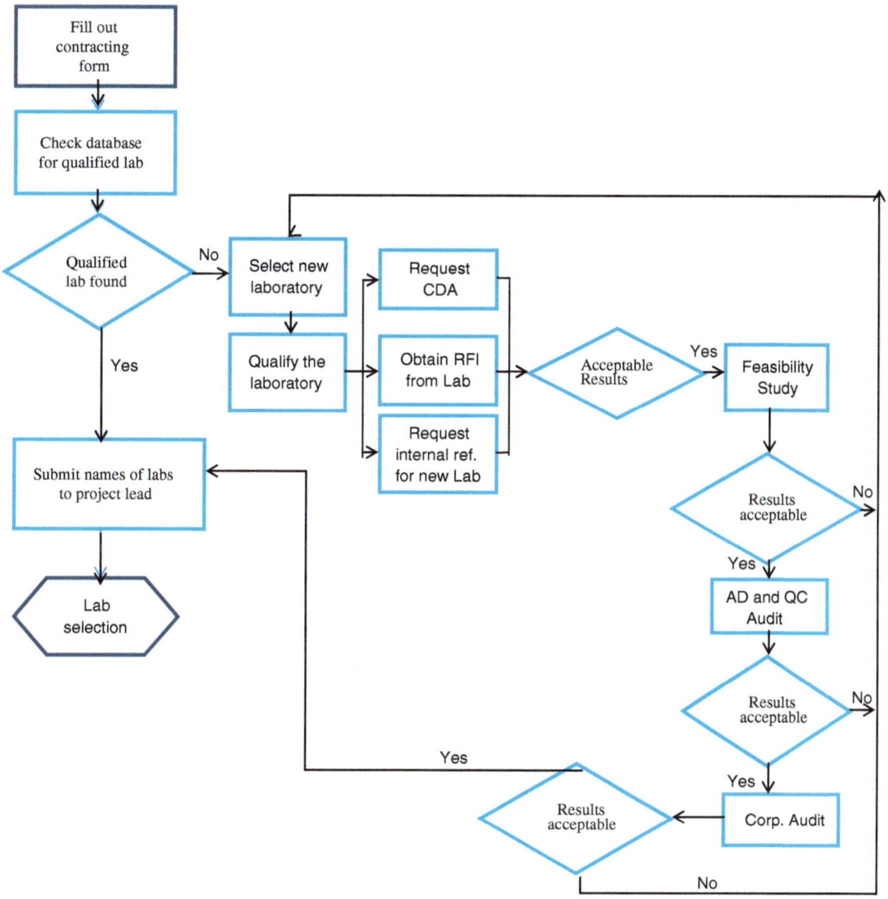

Fig. 5.2 Laboratory selection and qualification

A critical activity is the selection and qualification of the laboratory to perform the outsourcing testing. Figure 5.2 describes the process for laboratory selection and qualification.

Once the project is authorized for contracting and the contracting laboratory is selected, a contacting process is implemented. Figure 5.3 describes the process for contracting [3, 4].

Within the processes shown in Fig. 5.3 there are several important documents which must be prepared, such as, the Confidentiality Agreement (CDA), Request for Information (RFI), and Request for Quotation. Examples of the documents are shown in Figs. 5.4, 5.5, and 5.6.

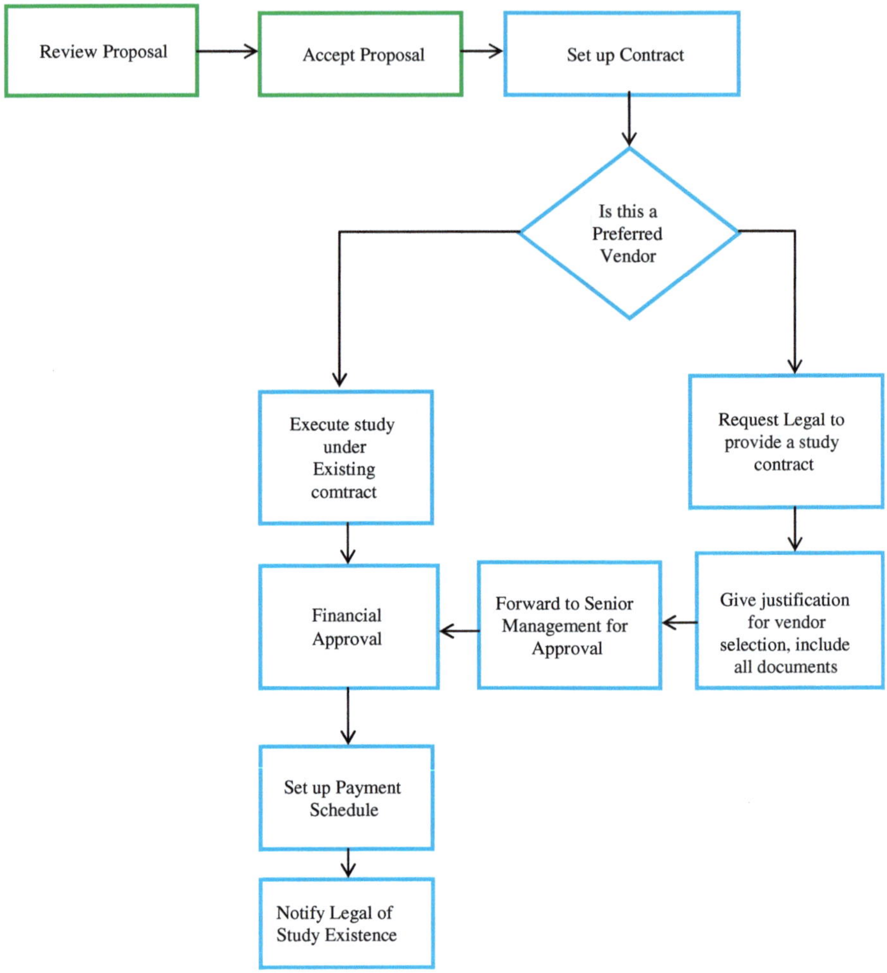

Fig. 5.3 Contracting process

5.2 Audit Form (Check List)

It is essential for the sponsor to establish that the laboratory providing support to the project is reliable and competent. An efficient and effective Audit Program is essential for the qualification of the outsourcing laboratory. There are two types of audits. One is an on-site audit, which should be performed for all new vendor laboratories. Two a paper audit which can be performed for approved vendor laboratories. The use of an audit form is a systematic approach to performing a complete and effective audit. Figure 5.7 demonstrates a generally used Audit Form.

Confidentiality Agreement

Representatives of **Sponsor Inc.**, having a mailing address havedisclosed or may disclose to Vendor, business information, technical information and/or ideas concerning **Sponsors**("Proprietary Information").

In consideration of any disclosure of Proprietary Information and for other consideration, the receipt of which is hereby acknowledged, you agree as follows:

1. You will hold in confidence and not possess or use except for purposes of discussions with **Sponsors** representatives (or as otherwise expressly authorized by **Sponsor**in writing) or disclose any Proprietary Information except information you can document (a) is in the public domain through no fault of yours, (b) was known to you prior to disclosure by **Sponsor**, or (c) was properly disclosed to you by another person without restriction. The foregoing does not grant you a license in or to any of the Proprietary Information. Notwithstanding the foregoing, specific aspects of Proprietary Information shall not be deemed to be within the foregoing exceptions when such exceptions apply only to more general knowledge or when the relevant specific aspects are identified using Proprietary Information disclosed under this Agreement.

2. If you are asked by **Sponsor** you will promptly return all Proprietary Information and all copies, extracts, notes and other objects or items in which it may be contained or embodied.

3. You will promptly notify **Sponsor** of any unauthorized release of Proprietary Information.

4. You acknowledge and agree that due to the unique nature of the Proprietary Information, any breach of this Agreement would cause irreparable harm to the **Sponsor** for which damages are not an adequate remedy and that **Sponsor** shall therefore be entitled to equitable relief in addition to all other remedies available at law.

5. The terms of this Agreement will remain in effect with respect to any particular Proprietary Information until you can document that it falls into one of the exceptions stated in Paragraph 1 above.

6. This Agreement is governed by the internal laws of the State of and may be modified or waived only in writing. If any provision is found to be unenforceable, such provision will be limited or deleted to the minimum extent necessary so that the remaining terms remain in full force and effect. You understand that this document does not obligate **Sponsor** to disclose any information or negotiate or enter into any agreement or relationship with you. The prevailing party in any dispute or legal action regarding the subject matter of this Agreement shall be entitled to recover attorneys' fees and costs.

Acknowledged and agreed on ____________, ____.

SPONSOR, INC
(Name) (Name of Vendor)

By: ______________________ By: ______________________

Name (Print) ______________________ Name (Print) ______________________

Title: ______________________ Title: ______________________

Fig. 5.4 Confidentiality agreement

1. Company information (General)
    a. Please provide information about your company's, Background, Services, Experience and Qualifications.
    b. Please provide a list of references for companies (minimum 5) for whom you provided analytical and/or stability support services, including contact names, addresses, and a brief description of the services provided.
    c. Do you provide preferred client relationships or strategic partnerships , if yes, please describe salient features of these relationships.
        i. Characterize your current business volume in the current analytical areas
            1. Method Development
            2. QC release testing
            3. Process Support
            4. Registration Stability Studies (FDA, ICH)
            5. Compendial testing
            6. Formulation support
            7. Cleaning validation/verification
            8. Sample storage and management
    d. Does your company possess any specialized expertise or analytical technology, if so please describe them?
    e. Describe your companies management structure and project Management processes, Please include an organizational chart and examples of project management work flow process.
    f. Do you perform feasibility studies and co-validations?
    g. Are you willing to engage in multi-year umbrella agreements, so that contracts are not need for every specific project? This would involve having standard terms and conditions in place with a pre-negotiated price structure.
    h. Do you offer cost savings based on volume of business, at what point and under what conditions the cost savings take effect.

2. Facilities
    a. Please describe the site facilities and a list of available equipment
    b. Please provide us with a list of most recent audits (last three years), with a redacted copy of one of your responses, as an example.
    c. Please provide job description for personnel at each level, and the average experience of the personnel.
    d. Describe your process for calibration and maintenance of the equipment
    e. Describe your document control and retrieval processes
    f. Describe your site safety program
    g. Describe your site security system
    h. Describe facility housekeeping and maintenance program

3. Regulatory Compliance
    a. Describe your Quality Assurance program
        i. Staffing (Number of personnel and qualifications)
        ii. Procedures to ensure compliance with guidelines and regulations for
            1. FDA
            2. Canada
            3. EMA
            4. Japan
        iii. What staff training is available, how is training documented?
        iv. Pleas furnish a copy of the Table of Contents for your Standard Operating Procedures (SOP)
    b. Please furnish copies of any forms FDA483, warning letters or consent degrees in the last three years, and of your two most recent site Audits

Fig. 5.5 Request for information

Quotation Request		
Requested	Items	Details
	Executive Summary of Project 1. Brief Introduction 2. Scope of Project	
	Cost for the Work to be Performed 1. Feasibility Studies 2. Method Development 3. Method Validation 4. Method Transfer 5. Excipient Analysis 6. Dosage Form Analysis 7. Chemical Analysis 8. Raw Material Analysis	
	Equipment Use Cost 1. HPLC,GC, Spectroscopy, Microscopy	
	Sample Preparation	
	Set-up Fees	
	Cost of Data Entry	
	Cost of Writing Reports	
	Cost of Re-Analysis	
	Cost of Storage of Samples	
	Cost of Disposal of Samples	
	Stability Study Fees 1. Set-up fees 2. Testing Fees 3. Management Fees	
	Cost for Auditing Technical Data and Compliance Monitoring	

Fig. 5.6 Quotation request

Audit Form

Instructions

1. Complete the Audit Checklist by checking either "Yes" or "No".
2. Identify all non-compliant items (No checked items)
3. Write Justifications for non-compliant Items
4. Correct all unjustified non-compliant items
5. Obtain signed report from vendor for all corrections and justifications
6. Attached signed vendor report to GMP audit form
7. Obtain vendor and client Signatures on GMP Audit For

Fig. 5.7 Audit form

SOP's and Operating Guidelines

	Yes	No
I. Specifications Development		
II. Documents for Submission to Regulatory Agencies		
III. Planned and Unplanned deviations		
IV. Notebook/Data Handling / Creation and Use		
V. Analysis Request/Sample Handling/Reports of analysis		
VI. Analytical Method Document		
VII. Performance Characteristics for Method Validation		
VIII. Method Validation Packages/Reports		
IX. Method Transfer Process		
X. Automated Instrument Implementation and Use		
XI. Personnel Training and Certification program		
XII. Rounding Rules		
XIII. Decimal Place/units Reporting for Analytical Data		
XIV. System Suitability for Chromatographic Methods		
XV. Replicate and Composite Size Determination for Dosage Form Assays		
XVI. Criteria for Identification and Qualification of Impurities		
XVII. Reporting Impurities, Including Degradation Products /Qualification of Chromatographic Peaks from Stability Samples as Degradation Product		
XVIII. Forced Degradation Studies for Method Development		
SOP's and Operating Guidelines(cont.)		
XIX. Laboratory Investigation of uncharacteristic analytical results		
XX. Equipment Calibration and Maintenance		
XXI. Decommissioning of Equipment		
XXII. GMP Material storage Management		
XXIII. Records Retention		
XXIV. Retention Sample Policy		

Fig. 5.7 (continued)

Item	Yes	No
Equipment Calibration and Maintenance		
I. Instrument calibration stickers		
II. Instrument use and maintenance log books		
III. Data system sample Log		
IV. Daily balance check log book		
V. HPLC/GC column use log book		
VI. Review/Approval documentation for vendor performed work		
VII. Instruments for GMP analysis clearly identified		
Chemicals/Reagents/Solutions:		
I. All chemical/reagents/solutions labeled:		
a. Date opened		
b. Date expired		
c. GMP or NON GMP use		
d. Special handling procedures		
II. Separated Acid and Bases storage bins		
III. Clean and Dirty glassware bins clearly identified		
Sample Receiving and Distribution system		
I. Numbering System for samples		
II. Storage of sample awaiting analysis		
III. Log book for sample tracking		
a. Submitter name		
b. Date received		
c. Tracking of sample movement		
d. Date dispensed for analysis(analyst name)		
e. Date of sample disposition (disposition)		
IV. Analytical Sample retention policy		
a. Sample retention log book		
i. Quantity of sample retained		
ii. Date of sample retention		
iii. Tracking of sample in and out of retention		
iv. Annual review of sample retain process		
v. Standardized report of analysis format		

Fig. 5.7 (continued)

GMP Material Storage Management:	Yes	No
I. Stability Chambers daily walk through to check on status, During off hours, a daily signing sheet should document time of walk through, name of personnel, and any discrepancies II. Use and maintenance log-book up to date III. GMP retains should be stored in a isolated chamber and IV. appropriately labeled a. GMP retention log book i. Quantity of sample retained ii. Date of sample retention iii. Tracking of sample in and out of retention iv. Annual review of sample retain procedure V. All removal of GMP retains need the approval of the Director of QA or designee		
Reference Standard Certification Program		
I. Designation of a Reference Standard Manager II. Limited access to reference standards (Ref. Std. Mgr or designee) III. Reference Standard control numbering system IV. Reference Standard request form V. Reference standard tracking system(log book or computer) VI. Reference Standard Certification and Classification process VII. Reference Standard Classification usage VIII. Reference Standard Classification Testing and Criteria IX. Certification and Recertification Process X. Reference Standard Certificate of Analysis XI. Documentation for original material to be certified XII. Storage of Reference Standard Material (-20C or lower) XIII. Retirement of Reference Standard process a. Control of Retains XIV. Handling of Reference Standard Material during usage a. Aliquot distribution b. Return policy c. Storage		

Fig. 5.7 (continued)

GMP Stability Process

	Yes	No

I. Control of stability chambers (digi-lab, etc.)
II. Backup system for power failures
III. Personal notification process for system failures
IV. Designation of a stability manager
V. Controlled access to stability area/chambers (Stab. Mgr. or designee)
VI. Documented access to stability area/chambers
VII. Tracking of incoming and outgoing samples
VIII. Preparation of stability samples
  a. GMP facility
  b. Isolated from other laboratory activities
IX. Protocol generation
X. Stability results data base
XI. Stability report generation
XII. Statistical modeling of data
  a. Determination of retest date
  b. Determination of shelf life

Laboratory Notebooks and Raw Data Control

I. Designation of notebooks procedure
II. Table contents
III. Table of abbreviations
IV. Formatting of notebook write up
  a. Required information
  b. Legible writing
  c. Use of worksheets (if applicable)
V. Review and Approval on each page
VI. Referencing from page to continuation pages
VII. Notebook review schedule
  a. Data review within time stated in SOP (usually 24 hours)
  b. GMP review within time stated in SOP(usually 3 months)
VIII. Adjunctive Notebook
  a. Storage of raw data
  b. Individual pages numbered and signed by analyst and reviewer
  c. Cross referenced to Notebook number and page numbers
IX. Notebooks retention policy

Fig. 5.7 (continued)

Personnel Training/Certification Program

	Yes	No
I. Training records up to date		
II. A documented analyst training/certification program should exist		
a. Analyst trained on all analytical activities regardless of education or experience		
b. All GMP analysis must be done by a trained analyst on that technique		
c. Retraining schedule should be specified		
d. Trainers for each technique should specified		
e. Trainers should attend a train the trainer course		
III. A List of required training for each job description should exist		
DATA Systems		
I. Documented Validated Data Systems		
a. Validation meets CFR21 part11 criteria		
b. Scheduled abbreviated re-validation (every 1 or 2 years)		
c. Detailed change control procedure		
d. All changes validated or qualified		
e. Authorized security administrator required for changes to system		
f. All data required approval before released to customers		

Approval Signatures

Vendor Signature_____________________________, Date______________

____

(Printed Name)

Client Signature_____________________________, Date______________

____

(Printed Name)

Fig. 5.7 (continued)

References

1. Hillstrom LC (2013) Outsourcing, reference for business, encyclopedia of business 2nd edn
2. Scully E (1995) Many factors to weigh in decision to outsource, National Underwriters Life and Health, financial services edition, p 10
3. Greco J (1997) Outsourcing: the new partnership. J Bus Strateg 18(4):48
4. Knowledge Encyclopedia (2008) Outsourcing, knowledge encyclopedia, Wikipedia

Chapter 6
Documentation

Abstract Documentation is an essential component in a GMP environment. Without proper documentation, Justification of data generation and data interpretation cannot be obtained. The old expression "If it's not Documented it's not Done" is well suited in a GMP/Regulatory environment. Included in this chapter are templates for several of the important documents that must available in an Analytical Department. Those described herein are Sample Submission, Technical Reports, Reports of Analysis, and Training Records.

6.1 Sample Submission

The sample submission process is the first interaction with the sample by the GMP Analytical Chemistry Department. Therefore a complete set of information must be obtained and tracked with the sample throughout the life of the sample. The sample submission process begins with the submitter completing the sample submission form (Fig. 6.1) and brings the complete sample submission form and the sample to the sample submission clerk. The clerk obtains two tracking tickets with the same number and places one on the sample and one on the sample submission form. The form is placed in a specified location at the sample submission area and the sample placed in an appropriate labeled container and stored under the proper conditions. The Stability clerk obtains a sample Identification number from the **Laboratory Information Management System (LIMS)**, this can also be done manually is a LIMS is not available. The clerk then enters the sample ID number on the sample submission form and on the sample container. The sample submission form is shown below in Fig. 6.1 [1].

The LIMS identifies samples waiting for analysis to the **Analytical Department (AD)** personnel. The AD analyst removes the sample submission form from the sample submission area and locates the sample container at the appropriate storage condition. The Analyst will than compare the tracking numbers and LIMS

T. Catalano, *Essential Elements for a GMP Analytical Chemistry Department*,
DOI: 10.1007/978-1-4614-7642-9_6,

Sample Submission Form		
Company Name	Sample submission tracking sticker placed here	
Submission ID Numbers		
Material description	Item Number	Project code
Submitters Name	Date Submitted	
Phone Number	Department Making the Submission	Location
Lot # and other information		
Storage Conditions __Ambient __Refrigerate ___Desiccate ___Freezer ___Protect from Light ___Other____		
Sample Objective ___R&D Release ___Formal Stability Testing ___Reference Standard ___Tox. Support ___Formulation Support ___Chemical Development Support ___Method Development ___Method Validation ___Formulation Screening ___Other______________________		
Testing Requested Appearance Dissolution Assay and Degradation Products Others___________________		
Sample Retriever	Date Retrieved	

Fig. 6.1 Sample submission form

identification number on the submission sheet and the sample container. If all numbers match correctly the analyst retrieves the sample and signs and dates the submission form. The analyst than notifies the submitter of his possession of the sample and gives the submitter the tracking number and LIMS identification number for information purposes. The Analyst upon retrieval of the samples performs the analysis as specified in the submission form, utilizing the appropriate methods. Upon completion of the analysis the analyst enter the data, the methods utilized, laboratory notebook reference, and any other appropriate information into the LIMS. The analyst then reviews the notebook and LIMS analytical report and signs off as analyst reviewer. The supervisor also reviews the notebook and the Analytical report and if no corrections are needed to be made, signs off as the approver of notebook entries and the analytical report. If corrections are needed the supervisor ensures that the corrections are made in the notebook and analytical report prior to approving the information. The supervisor in collaboration with other laboratory personnel, routinely inspect the sample submission area for samples no longer required by the analytical laboratory and ether holds samples as retention samples, as per the retention sample policy, or submits the samples for disposal. The sample submission form is shown below in Fig. 6.1 [1].

6.2 Technical Reports [2–4]

6.2.1 Development Reports

Development reports are among the most critical reports written by the Analytical Development Department. These reports are utilized to justify the data and conclusions which are included in the regulatory submission package. An example of a development report template is shown in Fig. 6.2.

6.2.2 Method Validation Report

The GMP's require that all methods must be successfully validated to support GMP studies required for submission to regulatory agencies. The Method validation report is prepared to support that the method is acceptable for use in a GMP environment. The Validation Report should contain or reference the experiments and data for the required validation parameters. Figure 6.3 is an example of a Validation Report Template.

6.2.3 Analytical Development Report

The analytical development report is description of the strategies, scientific logic, experiments evaluated, and the data collected to justify that all the developed analytical methodologies which will be utilized by the analytical department to support GMP analyses is consistent with current technology and is appropriate for its intended use. Figure 6.4 describes the information content which should be included in the analytical development report.

Development Report
Study Name

Report Number
Version, Date
Page Number

Report Number

Report Title

Author

Report Number
Study Name
Version, Date
Page Number

Prepared by: ____________________ ____________
Date

QC and Reviewed by: ____________________ ____________
Date

Reviewed by: ____________________ ____________
Date

Reviewed by: ____________________ ____________
Date

Approved by: ____________________ ____________
Date

Fig. 6.2 Development report template

Study Name

Report Number
Version, Date
Page Number

TABLE OF CONTENTS

Sections **Page No.#**

Study Name

Report Number
Version, Date
Page Number

1 Purpose & Objective

2 Summary

3 Discussion
 a Supportive data
 b Tables
 c Figures

4 Conclusions

5 References

Fig. 6.2 (continued)

Method Validation Report

Report Title

Project Number

Report Number

Author

Date

Report Approval Page

Author	Date:
Reviewer	Date:
Approval	Date:

Fig. 6.3 Validation report

Table of Contents

Validation Report

1. SUMMARY

2. EXPERIMENTAL

2.1 Materials

Standard:

Columns:

Equipment:

2.2 Sample Preparation

2.3 Placebo Formulations and Manufacture

3. RESULTS AND DISCUSSION

3.1 System Suitability

3.2 Accuracy/Precision at Assay Range

3.3 Intermediate Precision

3.4 Linearity

3.5 Specificity

Fig. 6.3 (continued)

.6 Limit of Quantitation (LOQ)

.7 Limit of Detection (LOD)

.8 Robustness

4 CONCLUSIONS

4 REFERENCES
5.1 Validation Protocol
5.2 ICH Q2(R1)
5.3 USP__NF__ <Pg>
5.4 Method Document
5.5 Laboratory Notebook Number, Pages

Fig. 6.3 (continued)

6.2.4 Stability Report

The stability report is a complete description and discussion of the stability Studies performed for submission to regulatory agencies. The report should include both primary studies and supportive studies, it should include justification for protocols, container closures, conditions used, timing for pull points, number of samples, sample positioning in the chambers, the analytical methodology utilized, description of the material (Drug Product or Drug Substance) and the identification and monitoring system for the stability chambers. Figure 6.5 demonstrates the content required in a stability report.

6.3 Analysis Report

All of the data generated from analytical testing is included in the Analysis Report, along with comments and conclusions. Figure 6.6 is an example of an Analysis Report.

6.4 Records

The Document Control portion of the organization is responsible for managing all records for the departments. This is a critical function, since the first stop for most regulatory audits is the review of personnel records. The individual Training

Analytical Development Report

Report Title

Project Number

Report Number

Author

Date

Approval Page

____

Author Date:

____

Reviewer Date:

____

Approval Date:

Fig. 6.4 Analytical development report

Table of Contents

Sections **Page No.#**

Analytical Development Report

1. Background

2. Physical and Chemical Properties of Drug Substance

3. HPLC Method Development (Drug Substance)

3.1. Initial Evaluation
3.2. Reverse-Phase Chromatography
3.3. Normal Phase Chromatography Exploration
3.4. Optimization of Resolution between Impurities.
3.5. Method Robustness

4. HPLC Method Development (Drug Product)

4.1 Initial Evaluation
4.2. Normal Phase Chromatography Exploration
4.3. Optimization of Resolution between Degradation Products.
4.4. Method Robustness

5. Method Development (Other Technologies)

5.1 Residual Sovents
5.2 Chirality
5.3 Particle Size
5.4 Polymorphorism
5.5 Excipients

Fig. 6.4 (continued)

6. Conclusion

7. References

1. Protocol, Method Validation: Drug Substance
2. Protocol, Method Robustness:
3. Protocol, Method Validation: Drug Product
4. Protocol, Method Robustness: Drug Product
5. Protocol, Method Validation: Other Technologies
6. Literature Articles
7. ICH Q2(R1)
8. USP__NF__ <Pg>
9. Method Documents
10. Laboratory Notebook Number, Pages

Fig. 6.4 (continued)

Records is a document that is almost always chosen for review by the regulatory agencies. Training records include individuals Position Description, Curriculum Vitae, and training documentation (SOP's, Guidelines, GMP/GLP compliance, Analytical Methodology, courses and seminars attended, etc.) and Professional Licenses or Certificates [5, 6]. A designated Compliance and Training Specialist is responsible for maintaining and updating these records, usually on an annual basis. An example of a training record document is shown in Fig. 6.7.

Stability Report

Report Title

Stability study ID:
Project Code:
Author:
Date

Approval Page

Author: ____Date_________

Reviewer: ____Date________

Approval: ____Date________

Fig. 6.5 Stability report

Stability Report

Table of Contents

Fig. 6.5 (continued)

Stability Report

1. Summary
   1.1 Purpose of the stability studies
   1.2 Description of material (Drug Substance/ Drug Product)
   1.3 Equipment (Stability Chambers, Stability Data System, etc.)
   1.4 Proposed Shelf Life
   1.5 Guidelines followed
   1.6 Stability Study Protocols
   1.7 Statistical Analysis Utilized

2. Lots used in the stability studies
   2.1 Lots for primary studies (Registration Studies)
   2.2 Lots for Registration like Studies (Phase III studies)
   2.3 Lots for supportive studies (Phase II studies)
   2.4 lots for scale up and Optimization studies (Development lots)

3. Summary of Stability Data
   3.1 Completed data tables for all studies
   3.2 Comparison of all data to specification criteria

4. Evaluation of stability data
   4.1 Identification of data which failed to meet criteria or show negative Trends
   4.2 Individual lot Regression Analysis
   4.3 Pooled lot Regression Analysis

5. Photo-stability studies
   5.1 Completed data tables
   5.2 Comparison of all data to specification criteria

6. Stability Conclusion
   6.1 Justification for proposed Shelf Life
   6.2 Statistical Analysis utilized
   6.3 Discussion of any nonconformity to specifications

Fig. 6.5 (continued)

Company Name Address	Analysis Report Submission ID: Lot #:________ Material Description:	Database: Approved: Y__ N__ Status:_________

Submission Details

Analysis Goal: (Release, Dev. Support, etc.)
Compound Number
Project Code
Requestors Name:

Sample Details

Sample ID: User ID:

Results Details

Test Task (eg. Appearance, Assay, etc.)		Status:_________	
		Approved:	Y___ N__
Component	Results	Version	Replicate
Method Number		No.#	
Analyst Name			
Notebook Number			
Notebook Page			
Instrument Number			
Compound Name			
Methodology	(HPLC, GC, etc.)		
Units	(%, mg/ml, ppm, etc.)		
Test -1			(1, 2, etc.)
Test-2			
Test-3			
Test-N			(1, 2, etc.)
Test- Mean			

Test- %RSD

Reviewing Analyst Signature: Date:
Printed Name:
Approving Supervisor Signature: Date:
Printed Name:

Fig. 6.6 Analysis report

Individual Training Record

Name Date

Title

Curriculum Vitae:

Professional License or Certificate Date Issued

Compliance Training

SOP's	2011	2012	2013	2014	2015
PD-10X	√	√	√		
PD- 20X	√	√	√		
Guidelines					
GL -10X	√	√	√		
GL – 20X	√	√	√		
GMP/GLP	√	√	√		

Courses and Seminars

Title Date Taken

Signature (Staff Member)______________________Date_____________

Signature (Compliance Specialist)_______________Date_____________

Fig. 6.7 Training record

References

1. 21 CFR Part 11.com Links to FDA Documentation
2. FDA Good Guidance Practices (2000) Federal Regulations, vol 65, no 182 pp. 56468–56480
3. Documents & Resources for Small Businesses & Professionals (2013), docstoc.com/Pharmaceutical Documents
4. Good Documentation Practice (2013), Wikipedia Encyclopedia
5. Faulkner B (2013) Documentation habits in the pharmaceutical industry, eHow.com/Pharmaceutical Documents
6. FDA Guidance, ICH Q10, Quality Systems

Chapter 7
Job Descriptions

Abstract Within an Analytical Department there are many different tracts of responsibility that exist. Tracts such as, Stability Management and analysis, Routine Testing and Validation, Scientist tracts and Team Leader tracts may exist. Each of these tracts must be clearly described along with the levels within the tracts, so that personnel have a clear vision for a potential career path. Care must be taken so that each tract has a minimum overlap so that there is no confusion as to the responsibility within each tract. A viable career tract is described along with the competencies required.

7.1 Career Tracks

In a GMP Analytical Chemistry Department it is important to define the different career tracts that are available. It is very important that each career tract is adequately described so that there is no confusion about job responsibilities between the different levels within a tract and between tracts. General areas of competency are described within each tract are as follows, however the depth, knowledge, performance requirements will increase proportionately with tract level and each level within the tracts.

1. Job Scope
2. Key result Areas

- Required Technical Skills
- Proficiency Skills
- Team Involvement
- Customer Interaction
- Level of Independence

T. Catalano, *Essential Elements for a GMP Analytical Chemistry Department*,

DOI: 10.1007/978-1-4614-7642-9_7,

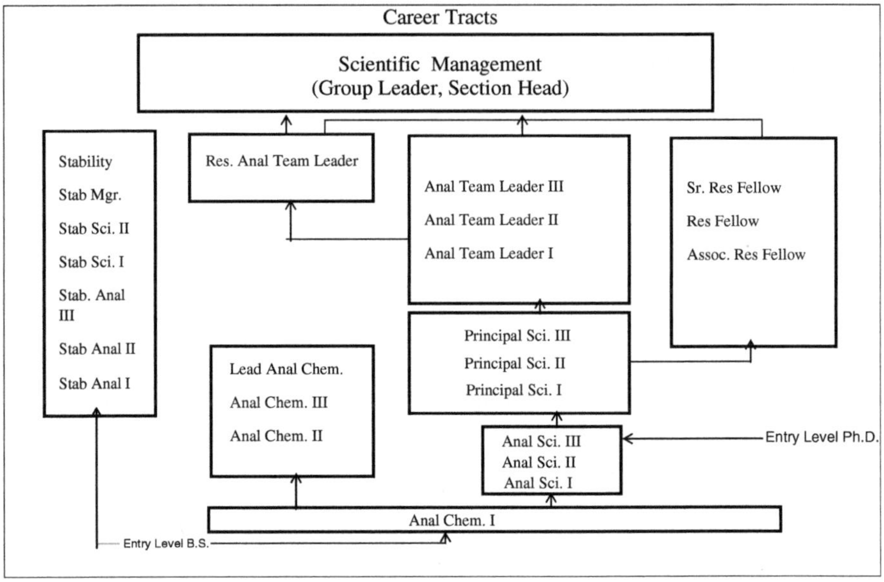

Fig. 7.1 Analytical department career tracts

3. Flexibility
   - Understands/Support Change
   - Capable of learning new skills due to change
4. Behavior Patterns
   - Results orientated
   - Continuous Learning
   - Seeks Feedback for improvement
   - Makes positive changes based on feedback
   - Motivation of others
   - Team Skills
5. Problem Solving and decisiveness
   - Applies relevant information to problem solving
   - Proactively Plans work steps
6. Impact and Influence
   - Behaviors and Performance Impact Results
7. Process Management
   - Effectively applies existing processes
   - Participates in the collection of metrics
   - Provides feedback to improve process

8. Communication Skills

   - Demonstrates adequate verbal skills, written skills, and demonstrates computer literacy

9. Development and Growth

   - Improves skills and proficiency towards next level
   - Greater Independence

Entry level B.S. Chemist will generally start as an Analytical Chemist I or a Stab Analyst I. After approximately one year Analytical Chemist I are moved to the analytical chemist II track or Analytical Scientist I tract depending on their performance and preference. Improvement in the above areas of competency can result in movement to higher levels within a career tract in a two or three years' timeframe however, movement between tracts is much more difficult and timely. Research Fellows can also be Analytical Team leaders and would have the Title as Research Analytical Team leader. An example of an Organizational Career Tract that can be applied to a GMP Analytical Chemistry Department is shown in Fig. 7.1.

Chapter 8
Motivating Personnel

Abstract A defined Career path is an excellent way to motivate personnel, so that each individual knows what they must accomplish to climb the career ladder. It also allows for lateral movement within the department so that the department does not become top heavy by utilizing promotions as the only avenue for growth. Another element for motivation is to consider a mentoring program. This is very well received, especially by younger staff members, because it allows the opportunity for professional growth without having to wait for the opportunity to come across their path during the course of regular business. Included in this chapter is a promotion process flow diagram and promotion template which will project a clear understanding of the process and what must be accomplished. In addition an example of a mentoring program is outlined which demonstrate the activities and discussions that should be considered.

8.1 Career Path

Motivating personnel involves several factors, one is have a clear and documented understanding of the organization career tracts as described in Chap. 7. The career tract acts as a road map for personnel to understand the opportunities afforded to them and the requirements needed to achieve them. The Career tract also demonstrates not only the ability to up within a career tract, but also latterly across career tracts which reduces the problem of the department becoming top heavy by utilizing promotions as the only avenue for growth. Having a clearly document promotion process is also needed to support positive motivation of the staff. Knowing the department process for promotions and clear documented requirements at each level, allows for non-subjective decisions to be made. Figures 8.1 and 8.2 are examples of a viable promotion process.

Another element for motivation is to consider a mentoring program. This is very well received, especially by younger staff members, because it allows the

T. Catalano, *Essential Elements for a GMP Analytical Chemistry Department*,
DOI: 10.1007/978-1-4614-7642-9_8,

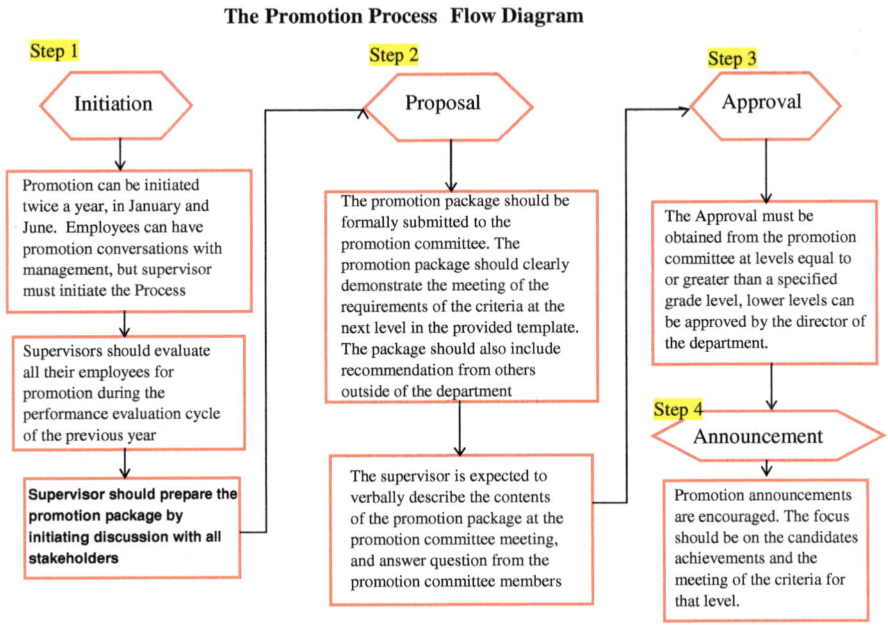

Fig. 8.1 The promotion process flow diagram

Promotion Process Template

Level	Skills/Expertise	Independence	Problem Solving	Interfaces	Communication	Accountability	Influence	Behavior
1	Details appropriate for each level	Details appropriate for each level	Details appropriate for each level	Details appropriate for each level	Details appropriate for each level	Details appropriate for each level	Details appropriate for each level	Details appropriate for each level
2								
3								
N								

Fig. 8.2 Promotion process template

opportunity for professional growth without having to wait for the opportunity to come across their path during the course of regular business. The mentoring program is usually required for entry level Ph.D.'s, but can be utilized by staff members at other levels on a voluntary basis. The mentoring program is described in Fig. 8.3.

Mentoring Program

Name of Mentor Candidate Start Date
Candidate's Title Department

Name of Mentor
Mentor's Title Department

I.<u>Candidate and Mentor Discussions</u>
These discussions are for both the candidate and the Mentor getting a more thorough understanding of each other's background and expertise

II.<u>Discussion of Research Projects</u>
The candidate and mentor discuss potential research projects that are of interest to the candidate, but are relevant to the Department needs and/or useful for the candidate's growth.

III.<u>Discussion of Research Project Goals</u>
The project chosen should have a completion date of approximately one year. Projects having longer completion dates have a tendency to lose momentum and are not as effective in being beneficial to the candidate. The Project goals should follow the format of S.M.A. R. T goals as shown below:

Short
Manageable
Attainable
Relevant
Tractable

Both the candidate and mentor should have reached consensus on the finalized goals.

Fig. 8.3 Mentoring program

A project plan should be constructed to identify all activities, their start dates and end dates and precursor links to dependent activities.

IV.Periodic Review of the Research Project.

Periodic reviews of the project status, usually six times a year, should be conducted between the candidate and the mentor. The function of these reviews is to:

- Review Data
- Obtain input from the mentor
- Consider changes in strategy or in goals (data driven)
- Determine if project is on track with the project plan
- Any financial concerns

V.Finalized Report

The candidate will produce a finalized report of the research adequate for presentation as an in-house seminar However, if deemed by the mentor to have the sufficient depth and originality; a paper will be prepared for publication to a peer review journal.

VI.Mentors Review of Candidate

The mentor will prepare a report on the candidate's performance, which will be sent to the candidate's supervisor and entered into the candidates personnel file.

Fig. 8.3 (continued)

After the completion of the mentoring cycle the candidate should supply the process owner of the Mentoring Process a report outlining the likes, Concerns, Suggestions and potential improvements to the program. In addition the candidate should include in the report the beneficial gains he/she obtained from the program, if any.

Chapter 9
Hiring the Best

Abstract The essential element for maintaining a department growth with new talent and replacing talent that has left the organization is the ability to Hire the Best talent. However there is fierce competition for obtaining the best personnel. Therefore a well thought out and structured hiring process is the best solution. Within this chapter a structured hiring process describes in detail, important elements such as, establishing staffing philosophies, the concept of a Master Interview Guide which would contain all the relevant information, including sets of questions for each competency, proper behavior during the interviewing process, and the interpretation of candidate facial and body language responses. Also included are documents that aid in the collection of information to enable a comprehensive evaluation of the candidate.

9.1 Hiring the Best [1]

Hiring is an essential element within the Analytical Department to keep the department growing with new talent, retaining the best talent and replacing talent that has left the organization. In addition a department must have the ability to be flexible, diverse, and resilient in order to deal with the unperceived circumstances that can be encountered in the course of the everyday challenges. The Hiring of the best personnel is the solution to dealing with these concerns. However, there is a fierce competition for good talent among the various organizations. Therefore a hiring process should be in place which is ongoing and is constantly connected with colleges, universities and recruiters. Implementing searches for key talent only when positions become available usually results in settling rather than selecting. The hiring process should focus on staffing philosophies which include both Technical and Behavioral attributes. The detailed competencies for each category are shown below:

T. Catalano, *Essential Elements for a GMP Analytical Chemistry Department*,
DOI: 10.1007/978-1-4614-7642-9_9,

Preparation
1. Review job requirements and candidates resume
2. Use Master Guide to select questions for each Competency
Welcome Candidate
1.Set a pleasant environment
2. Explain Agenda
Questioning
1. Ask only legal Questions
2. Try to obtain specific examples from the candidate, when appropriate
3. Ask follow up questions to help clarify or gain information
4. Paraphrase question when needed
Listening
1. Maintain eye contact with candidate
2. Re -state candidates response to ensure understanding
Managing the Interview
1. Make interview feel like a discussion, not a test or an interrogation
2. Focus and re-direct the discussion , as necessary
3. Control time of interview
Recruiting
1.Capitalize on opportunity to sell the organization and department
2. Fully answer candidates questions

Fig. 9.1 Interviewer behavior preparation list

Technical

- Functional and Technical Skills.
- Application and Problem Solving.
- Planning and Organization.

Behavioral

- Continuous Improvement.
- Listening and Learning.
- Participative Management.
- Coaching of others for skill development.
- Transparency.

The backbone for **Hiring the Best** is the Interview process, which should enable the interviewers to ask questions directed to staffing philosophies, thus allowing for comprehensive evaluation of the candidate. The Development of a **Master Interview Guide** was found to very helpful for arranging a thorough and systematic interview session. The master interview guide contains prepared interview questions, around each competency, broken down into several groups, and each interviewer would be given different group of questions to ask. This enables that the same question are not repeated by different interviewers and that a

Interviewing Do's and Don'ts

Typical question you CAN Ask

1. This position requires XX% travel, would that be a problem
2. This position may require relocation, any concerns
3. Do you clearly understand the position description
4. Does the position require any special accommodation for you to perform the required tasks.

Typical questions you CAN NOT ask

1. If you have any doubt about whether you can ask a question DON'T
2. Avoid asking questions that Do Not related to the candidates ability toperform the Job function.
3. Do not ask question related to Age, Marital Status, Children, Religion,Nationality, or Political Preferences.

Fig. 9.2 Legal do's and don'ts

Candidate Evaluation				
Candidates Name				Position
Interviewers Name/ Title				Date
1. Rating Against Competencies				
Competency				*Ratings*
1. Functional/Technical Skills				
2. Application and Problem Solving				
3. Planning and Organization				
4. Continuous Improvement				
5. Listening and Learning				
6. Participative Management				
7. Coaching of others for skill development				
8. Transparency				
Rating Key: 3. Exceeds Qualifications 2. Meets Qualifications 1. Below Qualifications +/- = Slightly above or below rating (eg. 2^+ or 2^-)				
2 Other Questions to Consider				
1. Desire for the Job	Y	N	NA	
2. Relevant work experience	Y	N	NA	
3.Team oriented/ works well with others	Y	N	NA	
4. Results oriented	Y	N	NA	
5. Leadership Potential	Y	N	NA	
6. Trainable	Y	N	NA	
Other Comments:				
3. Recommendation for Hire				
☐ Yes Supporting Notes				☐ NO Supporting Notes

Fig. 9.3 Candidate evaluation form

Positive
1. Good functional and technical skills
2. Detailed oriented
3. Can deal with complexity comfortably
4. Makes sound decisions based on data, experience, and Judgment
5. Effective communication skills
6. Good leader and coach
7. Independent
8. Sets clear objectives
9. Overcomes road blocks
10. Empowers, other, people want to do their best
Negative
1. Lack of interest to new technologies
2. Consumed by details and not project focus
3. May seem too academic
4. May appear suborn not willing to negotiate or Compromise
5. Makes things overly complex
6. Disorganize
7. Poor time manager
8. Not open to suggestion and input
9. Overly controlling
10. Not seen as a team player
11. May not respect other disciplines
12. Appears Arrogant, impatient
13. Interrupts during conversation
14. Not a good team player
15. Hypercritical of others

Fig. 9.4 Notepad alert questions

broader area is explored thus allowing for a more thorough interview. Also included in the Master Guide are guidelines on how the interviewer should behave so to create a comfortable and professional environment for the candidate (See Fig. 9.1). In addition the manual contains hints on interpreting certain responses and facial expressions, how to verbalize your organizations selling points, and lastly but extremely important is the detailing of the legal do's and don'ts that every interviewer must be aware of (see Fig. 9.2). The next step in the Process is the rating and evaluation of the candidate. This is done by completing the **Candidate Rating and Evaluation Form** (see Fig. 9.3) which is used by all the interviewers thus allow allowing for a common format for review. The last activity before the acceptance or rejection of the candidate is the discussion at the candidate evaluation session. Each participating interview must prepare his/her discussion and rating which must be delivered at the candidate evaluation session. The master interview guide includes all forms required and many helpful hints on interpreting the information gathered from the interview sessions. Taking notes during the interview is highly recommended, since it reduces the loss of important information which may be significant for the candidate rating and evaluation. An efficient way to effective note taking is to have prepared questions on your notepad to alert you to capturing pertinent ideas. An example of the types of questions is described in Fig. 9.4.

Reference

1. Klein D (2001) Klein learning solutions, training and consultants

Appendix I
Project Plan Development Checklist

I. R&D Plan	**Comments**
a. Toxicology Plan	
b. Regulatory Plan	
c. Documentation Plan (CMC)	
d. Marketing Plan	
II. Chemical Plan	
a. Chemical Availability	
b. Production Schedule	
c. ID. Comm. Manufacturing Process (Date)	
d. Impurities Identification and Qualification	
e. Solid State Characterization	
f. Stability Plan	
g. Release Testing	
h. Specification Setting	
III. Dosage Form Plan	
a. Development of Dosage Form Phase I Phase II Phase III/commercial Additional DF development	
b. Specification Setting	
C. Release Testing	

T. Catalano, *Essential Elements for a GMP Analytical Chemistry Department*,
DOI: 10.1007/978-1-4614-7642-9,

IV. Analytical Technology Plan

a. Development and validation of Analytical Methods ☐

b. Prepare Technical Reports ☐

c. Support Specifications Setting ☐

d. Manage Stability Program ☐

e. Support Formulation and Process Chemistry ☐

f. Release Testing ☐

V. Registration Plan

a. Preparation of Documents (CMC) ☐

b. Finalize Reports ☐

c. Identify Submission Date ☐

d. Respond to Regulatory Responses ☐

VI. Technology Transfer

a. Chemical Transfer ☐

b. Dosage form Transfer ☐

c. Analytical Methods Transfer ☐

d. Create the Technology Transfer Team ☐

e. Materials Sourcing Plan ☐

Appendix II
Project Strategy Document

Executive summary

I. Dosage form

Dosage form type	Development timing	Resource commitment

II. Pharmaceutics/Pre-formulation

Activity	Development timing	Resource commitment

T. Catalano, *Essential Elements for a GMP Analytical Chemistry Department*, DOI: 10.1007/978-1-4614-7642-9,

III. Safety assessment plan

Tox. study	Study timing	Resource commitment

IV. Analytical support

Activity	Development/Testing timing	Resource commitment

V. Drug substance

Needs	Timing	Resource commitment

VI. Outsourcing

Activity	Timing	Resource commitment	Est. cost

VII. Documentation

Document	Timing	Responsibility

VIII. Technology transfer

Activity	Timing	Responsibility

IX. Stability

Study	Timing/Duration	Responsible site	Outsource Y/N

X. Regulatory meeting

Meetings	Timing resource commitment

XI. Monitoring the plan/Issues and concerns

Activity	Comments

Appendix III
Project Risk Assessment

Risk factors analytical development			
	Weightings (1–10)		
Risk factors	Early stage	Mid stage	Late stage
Method development	3	6	9
Method validation	2	9	10
Availability of lots (DS &DP)	4	8	8
PSA support	7	9	9
Release testing	3	8	10
QC release	3	8	10
Manufacturing sites involvement	2	7	10
Specifications	1	3	8
Methods for alternative DF	1	7	8
Method optimization	1	5	9
Orthogonal methods	1	6	8
Reference standard support	3	7	10
Identification and qualification of impurities and degradation products	1	6	9
ICH validation	1	4	10
Technology transfer	1	4	8
Documentation, methods, validation Pkgs, regulatory, etc	3	6	10
Development and technical reports	6	8	9
Outsourcing	2	7	10
Budget/resources/headcount	2	5	8
Others			

Weightings = 1 least Critical
= 10 Most Critical

T. Catalano, *Essential Elements for a GMP Analytical Chemistry Department*,
DOI: 10.1007/978-1-4614-7642-9,

Risk factors pharmaceutical development			
Risk factors	Weightings (1–10)		
	Early stage	Mid stage	Late stage
Chemical and biological characterization	7	8	10
Formulation characterization	2	8	10
Manufacturability	2	5	10
Excipient compatibility	3	7	10
Formulation development	1	4	10
Clinical plan	10	10	10
Packaging	1	5	10
Scale up and process optimization	1	2	10
Manufacturing site changes	1	4	10
Alternate dosage forms (Technology)	3	6	8
Product definition	1	3	10
Clinical supplies	1	10	10
QC release	1	10	10
Development stability	8	10	10
Commercial process optimization	1	3	10
Technology transfer	1	4	10
Process validation	1	5	10
Registration stability	1	5	10
Commercial stability	1	5	10
Documentation	10	10	10
Outsourcing	10	5	3
Budget	10	10	10
Other			

Weightings = 1 least Critical
= 10 Most Critical

Risk Evaluation Data Sheet			
Risk Factors	Weight (1-10) X	Rate (1-5) =	Risk Value

Rating = 1 least Critical
= 5 Most Critical

Total Risk Value (X) =

Calculation of Plan Risk Evaluation

Calculate the Risk Evaluation using the following equations		
1.	Total risk evaluation	$(\%R_t) = \frac{(x_c - x_{\min})}{x_{max} - x_{min}} \times 100$
2.	Risk change index	$(\%R_c) = \frac{(x_n - x_c)}{x_{max} - x_{min}} \times 100$
3.	Low risk plan evaluation	$(\%R_{XL}) = \frac{(x_L - x_{\min})}{x_{max} - x_{min}} \times 100$

Definitions
x_{max} Maximum total risk value
x_{min} Minimum total risk value
x_L Total risk value for determined low risk plan
x_c Total risk value for current plan
x_n Total risk value for new plan
$\%R_t$ Percent total risk of current plan
$\%R_c$ Percent risk change between plans
$\%R_{XL}$ Percent risk evaluation of determined low risk plan
$x_{max} - x_{min}$ The difference between the theoretical max. value and the theoretical min. value

About the Author

Dr. Tom Catalano's has 30 years in the Pharmaceutical Industry in various positions including the Global Director of Analytical R&D, with major companies such as G. D. Searle, Pharmacia and Pfizer, which led to a multitude of experience in providing analytical support to dosage forms and drug substances including biologicals. Additionally, he worked for a biopharmaceutical company, Theravance, as Sr. Director of Technical Operations.

He provided analytical support to a variety of dosage forms and drug substances, which included:

Tablets, Soft Capsules, Hard Capsules, Creams and Gels

Injectables (Suspensions, Solutions or Lyophilized)

Controlled Release Forms (Oral and Injectable)

Manufacture of small molecules, peptides and proteins

In Dr. Catalano's 30 years in the industry, he made significant contributions to the development and registration of many products such as NutraSweet, Cytotec, Arthrotec, Celebrex, Bextra, Dynastat, Inspra and Vibativ (glycopeptide antibiotic).

Currently Dr. Catalano is president of a consulting company, PharmChem Analytical Consultants LLC.

T. Catalano, *Essential Elements for a GMP Analytical Chemistry Department*,
DOI: 10.1007/978-1-4614-7642-9,

Index

T. Catalano, *Essential Elements for a GMP Analytical Chemistry Department*,
DOI: 10.1007/978-1-4614-7642-9,

MIX
Papier aus verantwortungsvollen Quellen
Paper from responsible sources
FSC® C105338

If you have any concerns about our products,
you can contact us on
ProductSafety@springernature.com

In case Publisher is established outside the EU,
the EU authorized representative is:
Springer Nature Customer Service Center GmbH
Europaplatz 3, 69115 Heidelberg, Germany

Printed by Libri Plureos GmbH
in Hamburg, Germany